Electrospun Composite Nanofibers for Functional Applications

Electrospun Composite Nanofibers for Functional Applications

Editors

Ick-Soo Kim
Sana Ullah
Motahira Hashmi

MDPI • Basel • Beijing • Wuhan • Barcelona • Belgrade • Manchester • Tokyo • Cluj • Tianjin

Editors

Ick-Soo Kim
Textile Science and
Technology
Shinshu University
Ueda
Japan

Sana Ullah
Medicine Science and
Technology
Shinshu University
Ueda
Japan

Motahira Hashmi
Medicine Science and
Technology
Shinshu University
Ueda
Japan

Editorial Office
MDPI
St. Alban-Anlage 66
4052 Basel, Switzerland

This is a reprint of articles from the Special Issue published online in the open access journal *Polymers* (ISSN 2073-4360) (available at: www.mdpi.com/journal/polymers/special_issues/Electrospun_Composite_Nanofibers_Functiona_Applications).

For citation purposes, cite each article independently as indicated on the article page online and as indicated below:

LastName, A.A.; LastName, B.B.; LastName, C.C. Article Title. *Journal Name* **Year**, *Volume Number*, Page Range.

ISBN 978-3-0365-4524-0 (Hbk)
ISBN 978-3-0365-4523-3 (PDF)

Contents

Editorial

Electrospun Composite Nanofibers for Functional Applications

Sana Ullah [1,*], Motahira Hashmi [1,*] and Ick Soo Kim [1,2,*]

1 Graduate School of Medicine Science and Technology, Department of Science and Technology, Division of Smart Material, Shinshu University, Tokida 3-15-1, Ueda, Nagano 386-8567, Japan
2 Nano Fusion Technology Research Group, Institute for Fiber Engineering (IFES), Interdisciplinary Cluster for Cutting Edge Research (ICCER), Shinshu University, Tokida 3-15-1, Ueda, Nagano 386-8567, Japan
* Correspondence: sanamalik269@gmail.com (S.U.); motahirashah31@gmail.com (M.H.); kim@shinshu-u.ac.jp (I.S.K.)

Citation: Ullah, S.; Hashmi, M.; Kim, I.S. Electrospun Composite Nanofibers for Functional Applications. *Polymers* 2022, 14, 2290. https://doi.org/10.3390/polym14112290

Received: 25 May 2022
Accepted: 2 June 2022
Published: 5 June 2022

Publisher's Note: MDPI stays neutral with regard to jurisdictional claims in published maps and institutional affiliations.

Summary of the Special Issue:

Nanotechnology has deep roots in solving advanced and complex problems ranging from life sciences to water purification, from energy to structural applications, and from sensors to sustainable development. This Special Issue focuses on some of the applications of electrospun nanofibers in these areas. Our research group is working on applications of electrospun nanofibers in air filtration and facemasks [1,2], wound dressings [3–20], sensors [14], water purification [21], blood vessels [22], axons [23], adsorption [24,25], photosensitive materials [26], electronics [27,28], and various other fields of science and technology.

An interesting study was conducted on the fabrication of poly(ethylene-glycol 1,4-cyclohexane dimethylene-isosorbide-terephthalate) electrospun nanofiber mats (PEICT ENMs) for the potential infiltration of fibroblast cells. In this study, morphological, structural, and cytotoxicity assessments were performed. It was found that the PEICT nanofibers possessed excellent biocompatibility. Hence, the results confirmed that PEICT ENMs can potentially be utilized as a biomaterial [29]. Silver nanoclusters are also considered excellent antibacterial agents; another study in this Special Issue synthesized novel sericin-encapsulated silver nanoclusters (sericin-AgNCs) through a green synthesis route. Subsequently, these sericin-AgNCs were incorporated into ultrafine electrospun cellulose acetate (CA) fibers to assessing their antibacterial performance. It was confirmed via antibacterial activity testing that the AgNCs were sufficiently antibacterial in combination with cellulose acetate nanofibers to allow this nanofibrous web to be used as an antibacterial wound dressing or in similar applications. It was also observed that the addition of 0.17 mg/mL sericin-AgNCs to electrospun cellulose acetate fibers showed antibacterial activity of around 90%, while this value was increased to 99.9% by increasing the Ag nanocluster amount to 1.17 g/mL (i.e., increasing by 10 times) [30]. A comparison was drawn between polybutylene succinate (PBS) membranes and electrospun polyvinylidene fluoride (PVDF). Polymer fibers fabricated by rolling electrospinning (RE) and non-rolling electrospinning (NRE) showed different mechanical and morphological properties. Graphene oxide (GO) composite is neutral with regard to effects on mechanical properties. The PBS membrane offers a higher pore area than electrospun PVDF and can be used as a filter. The protein capture efficiency and protein staining were analyzed via the SPMA technique using albumin solution filtration. The fabricated membranes were compared with commercially available filters. RE with GO and PBS showed two times greater capture capacity than commercial membranes and more than sixfold protein binding as compared to the non-composite polymer. Protein staining results further verified the effectiveness of the fabricated membranes by showing a darker stain color [31]. In another research area, an in vitro cyst model was prepared in which hollow nanofiber spheres were developed, named "nanofiber-mâché balls." Electrospun nanofibers of a hollow shape were fabricated on alginate hydrogel beads. A fibrous geometry was provided by the balls having an inner volume of 230 mm^3. A route for nutrients and waste was developed by including two ducts.

This oriented migration was produced with a concentration gradient in which seeded cells attached to the inner surface, resulting in a well-oriented structure. Fibers attached to the internal surface between the duct and hollow ball restrict cell movement and do not allow them to exit the structure. An adepithelial layer on the inner surface was developed with this structure, and the nanofiber-mâché technique is most suitable for investigating cyst physiology [32]. The affinity of electrospun polyvinyl alcohol (PVA) nanofiber fabric was improved by modification with Cibacron Blue F3GA (CB). The bovine hemoglobin (BHb) adsorption capacity of the nanofiber fabrics at different concentrations of protein before and after treatment was studied in batch experiments. The original fibers possessed a BHb adsorption capacity of 58 mg/g, while the modified nanofiber fabric showed a capacity of 686 mg/g. The feed concentration and permeation rate were investigated. Static adsorption was performed to investigate the impact of the pH on BHb and bovine serum albumin (BSA) adsorption. Selective separation experiments were carried out at the optimal pH value. A reusability test was carried out by performing three adsorption–elution cycles [33]. Task-specific functionalized polymeric nanofiber mats were fabricated from homocysteine thiolactone and polyvinylpyrrolidone through electrospinning. This allows for post functionalization by a thiol–alkene "click" reaction. Modifications to electrospun nanofibers were made by introducing different functionalities under controlled conditions without damaging the nanostructure of the nanofibers. The results showed that modified nanofibers can be used for sensing and catalytic applications [34]. Recent trends in research showed that a large surface area, pore size, and geometry of nanofibers make them suitable for filtration applications. The filtration efficiency can be modified through nanonets of a smaller diameter than nanofibers. Polyacrylonitrile (PAN) was used to form nanonets, and their filtration efficiency was analyzed. Electrospun polyacrylonitrile acetyl methyl ammonium bromide (PAN–CTAB) possessed improved mechanical and thermal properties. PAN–CTAB nanofiber/nanonets showed 99% filtration efficiency as air filters and a low pressure drop of 7.7 mm H_2O at an air flow rate of 80 L/min. This study provides a new approach to the fabrication of air filters with higher filtration efficiency [35]. Another milestone in the field of skin care was achieved with the development of effective hydrophilic nanofibers (NFs) loaded with folic acid (FA). Electrospinning and electrospraying techniques were used. The morphological, thermal, mechanical, chemical, in vitro, and cytocompatibility properties were analyzed, and the results showed that the fabricated nanofiber mat has the potential for use as wound dressings and in tissue engineering applications [19]. A study on solid-state batteries (SSBs) gained attention for its efficiency in energy density and high-safety energy storage devices. Researchers have made various efforts to fill the gaps for thin solid-state-electrolytes (SSEs) with regard to their ionic conductivity, thermal stability, and mechanical strength. Composite polymer electrolyte (CPE)-reinforced PI nanofiber with succinonitrile-based solid composite electrolytes were developed to fill this gap. CPE showed a high ionic conductivity of 2.64×10^{-4} S cm^{-1} at room temperature. The developed material is fire resistant, is mechanically strong, and offers promising safety [36]. Research goals of controllable release and antibacterial properties were achieved via orange essential oil (OEO) and silver nanoparticles (AgNPs) deposited on a cellulose (CL) nanofiber mat. The fabrication of the finished nanofiber mats involved different steps like deacetylation and coating of silver nanoparticles prepared in OEO solution by an in situ method with two different concentrations. The successful deposition of AgNPs incorporated in the OEO was analyzed via SEM-EDS, TEM, XRD, and FT-IR. The tensile strength was recorded after each step of the treatment and compared with that of CA nanofibers. Well-treated nanofiber mats showed good antibacterial activity against Gram-positive and Gram-negative bacteria [37]. Researchers are keen to find high-quality nanomaterials for medical applications to change the future of medicine. The similarity between human tissues and electrospun nanofibers has opened new doors for researchers in the field of the medical applications of nanofibers. Until now, electrospun nanofibers have been restricted to tissue scaffolding applications, but a combination of nanoparticles with nanofibers could provide better function in photothermal, magnetic response, biosensing, antibacterial, and drug delivery applications. There are two methods to prepare hybrid nanofibers and nanoparticles (NNHs): electrospinning is the easy and simple way, and the other way is self-assembly. Both methods have been adopted to achieve drug release, antibacterial, and tissue engineering applications [38].

Author Contributions: Conceptualization, S.U. and M.H.; methodology, S.U.; validation, S.U., M.H. and I.S.K.; writing—original draft preparation, S.U. and M.H.; writing—review and editing, M.H. and I.S.K.; visualization, S.U. and I.S.K.; supervision, I.S.K.; project administration, S.U. All authors have read and agreed to the published version of the manuscript.

Funding: This research received no external funding.

Conflicts of Interest: The authors declare no conflict of interest regarding this editorial.

References

1. Hashmi, M.; Ullah, S.; Kim, I.S. Copper oxide (CuO) loaded polyacrylonitrile (PAN) nanofiber membranes for antimicrobial breath mask applications. *Curr. Res. Biotechnol.* **2019**, *1*, 1–10. [CrossRef]
2. Ullah, S.; Ullah, A.; Lee, J.; Jeong, Y.; Hashmi, M.; Zhu, C.; Joo, K., II; Cha, H.J.; Kim, I.S. Reusability Comparison of Melt-Blown vs Nanofiber Face Mask Filters for Use in the Coronavirus Pandemic. *ACS Appl. Nano Mater.* **2020**, *3*, 7231–7241. [CrossRef]
3. Ullah, S.; Hashmi, M.; Kharaghani, D.; Khan, M.Q.; Saito, Y.; Yamamoto, T.; Lee, J.; Kim, I.S. Antibacterial properties of in situ and surface functionalized impregnation of silver sulfadiazine in polyacrylonitrile nanofiber mats. *Int. J. Nanomed.* **2019**, *14*, 2693–2703. [CrossRef] [PubMed]
4. Ullah, A.; Saito, Y.; Ullah, S.; Haider, M.K.; Nawaz, H.; Duy-Nam, P.; Kharaghani, D.; Kim, I.S. Bioactive Sambong oil-loaded electrospun cellulose acetate nanofibers: Preparation, characterization, and in-vitro biocompatibility. *Int. J. Biol. Macromol.* **2021**, *166*, 1009–1021. [CrossRef]
5. Bie, X.; Khan, M.Q.; Ullah, A.; Ullah, S.; Kharaghani, D.; Phan, D.N.; Tamada, Y.; Kim, I.S. Fabrication and characterization of wound dressings containing gentamicin/silver for wounds in diabetes mellitus patients. *Mater. Res. Express* **2020**, *7*, 045004. [CrossRef]
6. Khan, M.Q.; Kharaghani, D.; Sanaullah; Shahzad, A.; Saito, Y.; Yamamoto, T.; Ogasawara, H.; Kim, I.S. Fabrication of antibacterial electrospun cellulose acetate/ silver-sulfadiazine nanofibers composites for wound dressings applications. *Polym. Test.* **2019**, *74*, 39–44. [CrossRef]
7. Khan, M.Q.; Kharaghani, D.; Sanaullah; Shahzad, A.; Duy, N.P.; Hasegawa, Y.; Azeemullah; Lee, J.; Kim, I.S. Fabrication of Antibacterial Nanofibers Composites by Functionalizing the Surface of Cellulose Acetate Nanofibers. *ChemistrySelect* **2020**, *5*, 1315–1321. [CrossRef]
8. Kharaghani, D.; Khan, M.Q.; Tamada, Y.; Ogasawara, H.; Inoue, Y.; Saito, Y.; Hashmi, M.; Kim, I.S. Fabrication of electrospun antibacterial PVA/Cs nanofibers loaded with CuNPs and AgNPs by an in-situ method. *Polym. Test.* **2018**, *72*, 315–321. [CrossRef]
9. Haider, M.K.; Sun, L.; Ullah, A.; Ullah, S.; Suzuki, Y.; Park, S.; Kato, Y.; Tamada, Y.; Kim, I.S. Polyacrylonitrile/ Carbon Black nanoparticle/ Nano-Hydroxyapatite (PAN/ nCB/ HA) composite nanofibrous matrix as a potential biomaterial scaffold for bone regenerative applications. *Mater. Today Commun.* **2021**, *27*, 102259. [CrossRef]
10. Ullah, S.; Hashmi, M.; Khan, M.Q.; Kharaghani, D.; Saito, Y.; Yamamoto, T.; Kim, I.S. Silver sulfadiazine loaded zein nanofiber mats as a novel wound dressing. *RSC Adv.* **2019**, *9*, 268–277. [CrossRef]
11. Ge, Y.; Tang, J.; Ullah, A.; Ullah, S.; Sarwar, M.N.; Kim, I.S. Sabina chinensis leaf extracted and in situ incorporated polycaprolactone/polyvinylpyrrolidone electrospun microfibers for antibacterial application. *RSC Adv.* **2021**, *11*, 18231–18240. [CrossRef] [PubMed]
12. Khan, M.Q.; Kharaghani, D.; Nishat, N.; Sanaullah; Shahzad, A.; Hussain, T.; Kim, K.O.; Kim, I.S. The fabrications and characterizations of antibacterial PVA/Cu nanofibers composite membranes by synthesis of Cu nanoparticles from solution reduction, nanofibers reduction and immersion methods. *Mater. Res. Express* **2019**, *6*, 075051. [CrossRef]
13. Kharaghani, D.; Gitigard, P.; Ohtani, H.; Kim, K.O.; Ullah, S.; Saito, Y.; Khan, M.Q.; Kim, I.S. Design and characterization of dual drug delivery based on in-situ assembled PVA/PAN core-shell nanofibers for wound dressing application. *Sci. Rep.* **2019**, *9*, 1–11. [CrossRef] [PubMed]
14. Kharaghani, D.; Suzuki, Y.; Gitigard, P.; Ullah, S.; Kim, I.S. Development and characterization of composite carbon nanofibers surface-coated with ZnO/Ag nanoparticle arrays for ammonia sensor application. *Mater. Today Commun.* **2020**, *24*, 101213. [CrossRef]
15. Hashmi, M.; Ullah, S.; Kim, I.S. Electrospun Momordica Charantia Incorporated Polyvinyl Alcohol (PVA) Nanofibers for Antibacterial Applications. *Mater. Today Commun.* **2020**, *24*, 101161. [CrossRef]
16. Sarwar, M.N.; Ali, H.G.; Ullah, S.; Yamashita, K.; Shahbaz, A.; Nisar, U.; Hashmi, M.; Kim, I.-S. Electrospun PVA/CuONPs/Bitter Gourd Nanofibers with Improved Cytocompatibility and Antibacterial Properties: Application as Antibacterial Wound Dressing. *Polymers* **2022**, *14*, 1361. [CrossRef]
17. Sarwar, M.N.; Ullah, A.; Haider, M.K.; Hussain, N.; Ullah, S.; Hashmi, M.; Khan, M.Q.; Kim, I.S. Evaluating Antibacterial Efficacy and Biocompatibility of PAN Nanofibers Loaded with Diclofenac Sodium Salt. *Polymers* **2021**, *13*, 510. [CrossRef]

18. Haider, M.K.; Ullah, A.; Sarwar, M.N.; Yamaguchi, T.; Wang, Q.; Ullah, S.; Park, S.; Kim, I.S. Fabricating antibacterial and antioxidant electrospun hydrophilic polyacrylonitrile nanofibers loaded with agnps by lignin-induced in-situ method. *Polymers* **2021**, *13*, 748. [CrossRef]
19. Parın, F.N.; Ullah, S.; Yıldırım, K.; Hashmi, M.; Kim, I.S. Fabrication and characterization of electrospun folic acid/hybrid fibers: In vitro controlled release study and cytocompatibility assays. *Polymers* **2021**, *13*, 3594. [CrossRef]
20. Hussain, N.; Ullah, S.; Sarwar, M.N.; Hashmi, M.; Khatri, M.; Yamaguchi, T.; Khatri, Z.; Kim, I.S. Fabrication and Characterization of Novel Antibacterial Ultrafine Nylon-6 Nanofibers Impregnated by Garlic Sour. *Fibers Polym.* **2020**, *21*, 2780–2787. [CrossRef]
21. Ullah, S.; Hashmi, M.; Hussain, N.; Ullah, A.; Sarwar, M.N.; Saito, Y.; Kim, S.H.; Kim, I.S. Stabilized nanofibers of polyvinyl alcohol (PVA) crosslinked by unique method for efficient removal of heavy metal ions. *J. Water Process Eng.* **2020**, *33*, 101111. [CrossRef]
22. Khan, M.Q.; Kharaghani, D.; Nishat, N.; Sanaullah; Shahzad, A.; Yamamoto, T.; Inoue, Y.; Kim, I.S. In vitro assessment of dual-network electrospun tubes from poly(1,4 cyclohexane dimethylene isosorbide terephthalate)/PVA hydrogel for blood vessel application. *J. Appl. Polym. Sci.* **2019**, *136*, 47222. [CrossRef]
23. Khan, M.Q.; Kharaghani, D.; Nishat, N.; Ishikawa, T.; Ullah, S.; Lee, H.; Khatri, Z.; Kim, I.S. The development of nanofiber tubes based on nanocomposites of polyvinylpyrrolidone incorporated gold nanoparticles as scaffolds for neuroscience application in axons. *Text. Res. J.* **2019**, *89*, 2713–2720. [CrossRef]
24. Hashmi, M.; Ullah, S.; Ullah, A.; Khan, M.Q.; Hussain, N.; Khatri, M.; Bie, X.; Lee, J.; Kim, I.S. An optimistic approach "from hydrophobic to super hydrophilic nanofibers" for enhanced absorption properties. *Polym. Test.* **2020**, *90*, 106683. [CrossRef]
25. Hashmi, M.; Ullah, S.; Ullah, A.; Akmal, M.; Saito, Y.; Hussain, N.; Ren, X.; Kim, I.S. Optimized Loading of Carboxymethyl Cellulose (CMC) in Tri-component Electrospun Nanofibers Having Uniform Morphology. *Polymers* **2020**, *12*, 2524. [CrossRef]
26. Khatri, M.; Ahmed, F.; Ali, S.; Mehdi, M.; Ullah, S.; Duy-Nam, P.; Khatri, Z.; Kim, I.S. Photosensitive nanofibers for data recording and erasing. *J. Text. Inst.* **2021**, *112*, 429–436. [CrossRef]
27. Hussain, N.; Mehdi, M.; Yousif, M.; Ali, A.; Ullah, S.; Siyal, S.H.; Hussain, T.; Kim, I.S. Synthesis of highly conductive electrospun recycled polyethylene terephthalate nanofibers using the electroless deposition method. *Nanomaterials* **2021**, *11*, 531. [CrossRef]
28. Hussain, N.; Yousif, M.; Ali, A.; Mehdi, M.; Ullah, S.; Ullah, A.; Mahar, F.K.; Kim, I.S. A facile approach to synthesize highly conductive electrospun aramid nanofibers via electroless deposition. *Mater. Chem. Phys.* **2020**, *255*, 123614. [CrossRef]
29. El-Ghazali, S.; Khatri, M.; Mehdi, M.; Kharaghani, D.; Tamada, Y.; Katagiri, A.; Kobayashi, S.; Kim, I.S. Fabrication of poly(Ethylene-glycol 1,4-cyclohexane dimethylene-isosorbide-terephthalate) electrospun nanofiber mats for potential infiltration of fibroblast cells. *Polymers* **2021**, *13*, 1245. [CrossRef]
30. Mehdi, M.; Qiu, H.; Dai, B.; Qureshi, R.F.; Hussain, S.; Yousif, M.; Gao, P.; Khatri, Z. Green synthesis and incorporation of sericin silver nanoclusters into electrospun ultrafine cellulose acetate fibers for anti-bacterial applications. *Polymers* **2021**, *13*, 1411. [CrossRef]
31. Sathirapongsasuti, N.; Panaksri, A.; Boonyagul, S.; Chutipongtanate, S.; Tanadchangsaeng, N. Electrospun Fibers of Polybutylene Succinate/Graphene Oxide Composite for Syringe-Push Protein Absorption Membrane. *Polymers* **2021**, *13*, 2042. [CrossRef] [PubMed]
32. Huang, W.Y.; Hashimoto, N.; Kitai, R.; Suye, S.I.; Fujita, S. Nanofiber-mâché hollow ball mimicking the three-dimensional structure of a cyst. *Polymers* **2021**, *13*, 2273. [CrossRef] [PubMed]
33. Liu, S.; Mukai, Y. Selective adsorption and separation of proteins by ligand-modified nanofiber fabric. *Polymers* **2021**, *13*, 2313. [CrossRef] [PubMed]
34. Valverde, D.; Muñoz, I.; García-Verdugo, E.; Altava, B.; Luis, S.V. Preparation of nanofibers mats derived from task-specific polymeric ionic liquid for sensing and catalytic applications. *Polymers* **2021**, *13*, 3110. [CrossRef] [PubMed]
35. Kang, H.K.; Oh, H.J.; Kim, J.Y.; Kim, H.Y.; Choi, Y.O. Performance of PAN–CTAB Nanofiber/Nanonet Web Combined with Meltblown Nonwoven. *Polymers* **2021**, *13*, 3591. [CrossRef]
36. Yuan, B.; Zhao, B.; Cong, Z.; Cheng, Z.; Wang, Q.; Lu, Y.; Han, X. A flexible, fireproof, composite polymer electrolyte reinforced by electrospun polyimide for room-temperature solid-state batteries. *Polymers* **2021**, *13*, 3622. [CrossRef]
37. Oil, O.E.; Nanoparticles, S.; Phan, D.; Khan, M.Q.; Nguyen, V.; Vu-manh, H.; Dao, A.; Thao, P.T.; Nguyen, N.; Le, V.; et al. Investigation of Mechanical, Chemical, and Antibacterial Properties of Electrospun Cellulose-Based Scaffolds Containing Orange Essential Oil and Silver Nanoparticles. *Polymers* **2022**, *14*, 85.
38. Zhang, M.; Song, W.; Tang, Y.; Xu, X.; Huang, Y.; Yu, D. Polymer-Based Nanofiber–Nanoparticle Hybrids and Their Medical Applications. *Polymers* **2022**, *14*, 351. [CrossRef]

Review

Polymer-Based Nanofiber–Nanoparticle Hybrids and Their Medical Applications

Mingxin Zhang [1], Wenliang Song [1,*], Yunxin Tang [1], Xizi Xu [1], Yingning Huang [1] and Dengguang Yu [1,2,*]

1   School of Materials and Chemistry, University of Shanghai for Science and Technology, Shanghai 200093, China; 203613006@st.usst.edu.cn (M.Z.); 213353179@st.usst.edu.cn (Y.T.); 192432632@st.usst.edu.cn (X.X.); 1935040104@st.usst.edu.cn (Y.H.)
2   Shanghai Engineering Technology Research Center for High-Performance Medical Device Materials, Shanghai 200093, China
*   Correspondence: wenliang@usst.edu.cn (W.S.); ydg017@usst.edu.cn (D.Y.)

Abstract: The search for higher-quality nanomaterials for medicinal applications continues. There are similarities between electrospun fibers and natural tissues. This property has enabled electrospun fibers to make significant progress in medical applications. However, electrospun fibers are limited to tissue scaffolding applications. When nanoparticles and nanofibers are combined, the composite material can perform more functions, such as photothermal, magnetic response, biosensing, antibacterial, drug delivery and biosensing. To prepare nanofiber and nanoparticle hybrids (NNHs), there are two primary ways. The electrospinning technology was used to produce NNHs in a single step. An alternate way is to use a self-assembly technique to create nanoparticles in fibers. This paper describes the creation of NNHs from routinely used biocompatible polymer composites. Single-step procedures and self-assembly methodologies are used to discuss the preparation of NNHs. It combines recent research discoveries to focus on the application of NNHs in drug release, antibacterial, and tissue engineering in the last two years.

Citation: Zhang, M.; Song, W.; Tang, Y.; Xu, X.; Huang, Y.; Yu, D. Polymer-Based Nanofiber–Nanoparticle Hybrids and Their Medical Applications. *Polymers* **2022**, *14*, 351. https://doi.org/10.3390/polym14020351

Academic Editors: Ick-Soo Kim, Motahira Hashmi and Sana Ullah

Received: 30 December 2021
Accepted: 14 January 2022
Published: 17 January 2022

Publisher's Note: MDPI stays neutral with regard to jurisdictional claims in published maps and institutional affiliations.

Keywords: polymer composites; nanoparticle; polymer blends; medical applications; electrospinning

1. Introduction

COVID-19 has contributed to a worldwide healthcare crisis resulting in several hundreds of thousands of deaths over recent years and complications that can lead to severe pneumonia [1,2]. In synergy with Polymer Engineering, Bioengineering, and Advanced Manufacturing, the mRNA vaccine was successfully developed and blocked the further spread of the virus on a large scale [3,4]. The mRNA-1273 vaccine approved for use under urgent circumstances has the potential to be unsafe for delivery, being inherently unstable while the large density of negative charges makes it difficult to cross cell membranes [5,6], and the latest research uses polymer-based composites as carriers for effective vaccine delivery [7].

In medical applications, polymers with superior biocompatibility are often used to not produce antibody reactions or even stimulate inflammation in contact with the human body [8–10]. Synthetic polymers include: polycaprolactone (PCL), poly (vinyl alcohol) (PVA), polylactide (PLA) and poly (lactic-co-glycolic acid) (PLGA) [11,12]; natural polymers include: Chitosan (polysaccharide), cellulose (polysaccharide), gelatin (protein), silk proteins (proteins) [13], etc. In biomedical applications, natural polymers are often perceived as relatively safe, biocompatible and degradable polymers [14]. Cellulose is a linear polysaccharide, the main component of the cell wall, which is currently the most used in biological applications from bacteria [15]. Taokaew et al. composite nanofibers based on bacterial cellulose with *Garcinia mangostana* peel extract exhibited potent bacterial inhibition against Gram-positive bacteria while treating the membranes with MCF-7 breast cancer cells for 48 h, only 5% of the cancer cells remained viable [16]. The protein class of silk fibroin, unlike other natural polymers, has outstanding elasticity and strength, which is of interest in tissue engineering applications [17,18].

The advantages of synthetic polymers over natural polymers are their stability, superior mechanical properties and degradability. PCL is an aliphatic semicrystalline polymer with hexanoic acid formed by esterolysis [19], which can be excreted through the digestive system, thus making it biocompatible, biodegradable, non-toxic and popular in medical applications, such as drug delivery, wound plug dressings and stents. The use of PCL scaffolds for tissue engineering has been reported in detail by Janmohammadi et al. [20]. Zavan and co-works prepared a bilayer fibrous scaffold with the outer layer of PCL exhibiting the best in vitro response to fibroblasts and weak adhesion to cellular tissues due to the hydrophobicity of PCL to cellular tissues, while the inner layer used gelatin-modified PCL to significantly promote gene expression in endothelial cells [21].

Nanoscale polymers have a wide range of uses in various fields, especially in medical applications, where the electrospinning process (Figure 1) is one of the most convenient ways to prepare nanofiber membranes (1–100 nm) directly and continuously [22–26]. Laboratory electrospinning setup generally has three components: a high voltage electrostatic generator (1–30 kv), a spinning head and a collector (opposite charge target) for the process of producing various continuous nanofiber assemblies with controlled morphology and dimensions from polymer solutions or melts under a high voltage electric field [27–29]. The development of the electrospinning (ES) process was a somewhat difficult journey. In the 16th century, William Gibert discovered a conical droplet formed when charged amber approached water droplets. Micro-droplets are also ejected from the conical droplets. The beginning of electrospinning technology is recognized to be in 1934, when Anton Formalas invented a device to prepare polymer fibers by electrostatic force action, presenting for the first time how a polymer solution could form the jet between electrodes. Subsequent years of intermittent patent publication have not attracted much attention from researchers. In 1990, Reneker's research team at the Acolon University, USA, went deeper into the electrospinning (ES) process and applications, with ES of various organic polymer solutions into fibers, ES process is rapidly becoming a research hotspot, entering a new era of vigorous development [30,31].

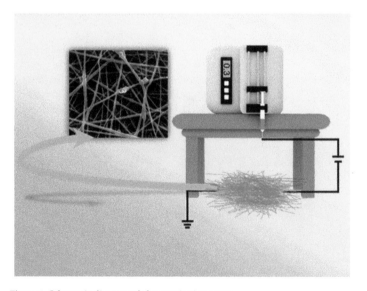

Figure 1. Schematic diagram of electrospinning setup.

To prepare the desired nanofibers, the electrospinning process usually requires setting parameters, the main influencing factors being the system (nature of the polymer, solution viscosity, conductivity and surface tension) [32], process parameters (voltage, receiving distance and flow rate) [33] and environmental factors (temperature and humidity) [34]. Specific effects on electrospun fiber morphology are shown in Table 1.

Table 1. Effect of control parameters on the morphology of electrospun fibers.

	Influence Factors	Influence Results	Reason for Influence	Ref.
System parameters	Polymer concentration	The higher the concentration, the coarser the fiber	As polymer concentration or molecular weight increases, so does solution viscosity. Greater entanglement between molecule chains and increased intermolecular Coulomb forces result from this condition. As a result, the fiber diameter expands.	[35]
	The molecular weight of polymers	The higher the molecular weight, the thicker the fiber		[36,37]
	Surface tension	The higher the surface tension, the finer the fiber	The droplet's surface tension rises, and the jet must expend more energy to offset this negative effect. The speed of the jet slows down, requiring more time to stretch the fibers. As a result, the fiber diameter decreases.	[38]
	Conductivity	Conductivity increases within a reasonable range; fiber diameter decreases and increases again; fiber diameter is not controllable	The charge accumulates on the surface of the jet when the conductivity is increased. The fibers stretch more quickly in this state. As a result, the diameter of the fiber is lowered. The coulombic repulsion at the jet interface is intensified when the solution conductivity is raised further. Uncontrollable fiber diameter distribution results from the unstable bending whip effect.	[39–41]
Process parameters	Voltage	The fiber diameter decreases with higher voltage and increases with higher voltage	As the voltage rises, the charge density on the jet's surface rises in accumulation. The circumstance may lead to a significant effect of jet stretching. Therefore, the fiber diameter decreases. The flow rate at the spinneret, on the other hand, increases as the voltage is raised more. Instead, the diameter of the fiber rises.	[42]
	Flow rate	The flow rate increases; the fiber diameter increases; and further increases may result in droplets	The solution at the spinneret rises as the flow rate increases. This condition causes the fiber's diameter to thicken. When the flow rate is too fast, the solution's gravity causes it to trickle straight down.	[43,44]
	Receiving distance	Acceptance distance enlarges and fiber diameter reduces	The additional receiving distance gives the jet more time to extend. The fiber's diameter shrinks in this circumstance.	[45]
Environmental factors	Temperature	Within a reasonable range, the fiber diameter decreases as the temperature increases	The temperature has the greatest influence on the viscosity of the solution. The viscosity of the solution reduces as the temperature rises. The intermolecular Coulomb force is lessened in this scenario.	[46]
	Humidity	Humidity increases and grooves appear on the fiber surface	When humidity is too high, fiber production is accelerated. Water droplet condensation on the fiber surface. Wrinkles occur on the surface of the fibers as a result of this process.	[47]

Synthetic polymers have modifiable mechanical properties, but the hydrophobic nature contributes to an absence of cell adhesion and the onset of inflammation, which is usually modified by natural polymers. One problem that most natural polymers suffer from is the deficiency of certain mechanical properties [48]. To combine the advantages of various polymers and overcome their limitations, the desired properties have been achieved using blends of these polymers instead of using single polymers by optimizing the ratio between the components of the blend. During drug release, the hydrophobic properties of the polymer seriously affect the release performance of the loaded drug. Huo and co-workers used PCL blended with gelatin to effectively improve the hydrophobic properties of the composite fiber. Drug release demonstrated that the increased PCL content made the composite fiber more hydrophobic, which belonged to the slow release of artemisinin (ART) and enhanced the therapeutic effect [49]. Chen et al. improved the defect of gelatin's poor hemostasis ability for large wounds by co-mixing sodium alginate, which can form a gel on the wound surface, and achieved hemostasis of rat wounds in only 1.53 min [50].

Nanofiber membranes are prepared by the ES process of polymers, and the application scenario is constrained to fiber scaffolds [51–53]. Researchers are not satisfied with this single role, so various functional nanofiber membranes have been prepared by ES in combination with multifunctional materials, among which NNHs show great potential to combine the advantages of nanoparticles with the properties of polymers. Additionally, the is a wide range of application scenarios [51–58]. Of interest is that as with decades of research, the number of electrospun nanoparticle articles published occupies half the number of electrostatic spinning (Figure 2). It shows the pivotal position of NNHs.

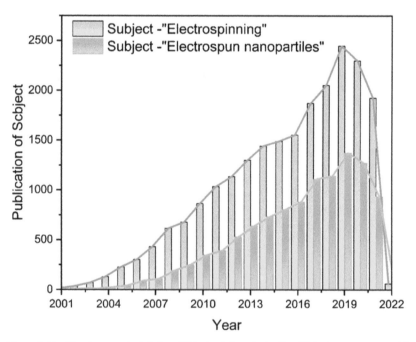

Figure 2. The literature search statistics of "electrostatic spinning" and "electrospun nanoparticles" on the "Web of Science" platform, respectively.

Multiple nanoparticles have been successfully prepared into nanofibers, which mainly include: metal [59], metal oxide [60] and polymer nanoparticles (NPs) [61,62]. The presence of NPs in polymer solution is uniformly dispersed and will be randomly distributed on the surface or inside the nanofibers (NFs). Therefore, the co-blending of NPs with polymer solutions is the most common form. During the process of electrospinning the mixture into fibers, the homogeneity of NPs in the polymer solution and the interfacial interaction between nanoparticles and polymer solution significantly affect the performance of NNHs. To attenuate the effects of the above factors, the NPs can be mixed by physical mixing methods (stirring and sonication), by dissolving the NPs and polymer in different solvents separately or adding a certain amount of surface activator. The electrospun fiber precursor working solution can effectively reduce the phenomenon of nanoparticle agglomeration and even blockage of the spinning heads after mixing uniformly [63–65]. Secondly, other processes combined with ES technology can also effectively load NPs into NFs. For instance, the modification of NFs by plasma technology. This methodology promotes the interaction between NPs and easily loads NPs on the nanofiber surface [66]. The electrospray [67] or magnetron sputtering [68] technology allows the uniform spraying of NPs on the fiber surface. The single-fluid ES process requires mixed solutions with a certain viscosity and dispersion. Otherwise, a non-spinnable or agglomerative phenomenon will occur. Multi-fluid dynamics have been studied for many years. The design of the spinneret is especially important. The morphological structure of the nanofibers is similar to the spinneret structure [69]. Multi-fluid dynamics ES processes often require only one fluid available for spinning to form nanofibers [70]. Zheng et al. used TiO_2 suspension as the sheath layer and PEO as the core layer, which perfectly avoided the enrichment of TiO_2NPs using a modified coaxial electrospinning process, and the uniform TiO_2NPs on the nanofiber surface enhanced the absorption of UV light [71]. The NPs are loaded in nanofibers based on the above single-step process, and if the NPs are not well dispersed in the working solution, many self-assembly strategies are available to grow NPs in the fibers by initiation effects in the precursor solution, including: in situ synthesis [72], hydrothermal-assistance [73] and calcination [74].

NNHs prepared by combining the characteristics of biocompatible polymers and multifunctional nanoparticles are already practical in medical applications [75,76]. The objective of this review is to investigate the preparation of nanofiber and nanoparticle hybrids in-depth by combining different processes with ES; concentrate on the most recent breakthroughs of NNHs in the medical area (Figure 3); and provide clear concepts for research workers in material preparation and application.

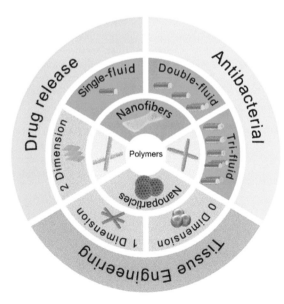

Figure 3. Electrospun fiber and NPs structures and their hybrids in medical direction.

2. The Methods for Creating Polymer-Based Nanofiber-Nanoparticle Hybrids (NNHs)

2.1. Overlapping of Electrospinning and Other Techniques

ES is a top-down molding process for the direct and continuous preparation of nanofibers. NNHs in the preparation of NPs in the spinning precursor solution, due to a certain stability, will not form a homogeneous solution in organic solvents. In addition to the large-scale manufacturing of NNHs, other technologies are often used in further combination with ES.

AgNPs in PCL nanofibers suffer from inhomogeneous dispersion and irregular ion release. To address this problem, Valerini and co-workers employed the magnetron sputtering technique to sputter Ag targets onto the surface of PCL nanofibers at a low power of 500 w. SEM images of PCL NFs after Ag magnetron sputtering show higher contrast. Composite NFs with 22 nm size AgNPs were deposited on the surface. Although the Ag content was only 0.1 wt.%, the PCL-Ag nanofibers showed a faster and stronger antibacterial effect by depositing all AgNPs on the nanofiber surface by magnetron sputtering technique [77]. Immersion of nanofibers (NFs) in NPs suspension loaded with NPs is the simplest approach; however, significant rejection occurs due to the different hydrophilic properties of NPs and NFs, leading to ineffective results. Liu et al. introduced oxygen-containing polar functional groups to the surface of PLLA membranes by a facile plasma technique. Negatively charged pPLLA membranes, impregnated in AgNPs solution, were well aggregated with NPs on the fiber surface driven by electrostatic interactions [78]. Electrospray, a sister technology to electrospun fibers, is commonly employed to generate NPs [79]. Fahimirad and co-workers sprayed chitosan NPs loaded with curcumin onto the surface of PCL/chitosan/curcumin nanofibers by electrospray, which showed no significant change in nanofiber diameter. While effectively promoting the swelling ability and degradation rate of nanofibers, the composite fibers healed 98.5% of the wounds infected by MRSA through wound closure experiments [80].

2.2. Encapsulating Nanoparticles in the Working Fluids

The presence of NPs necessarily affects the spinnability and composite fiber morphology during the preparation of NNHs. NPs appear agglomerated in the working solution, and needle blocking may occur during ES, or the NPs will not appear uniformly on the fibers. These issues can lead to reduced mechanical properties and loss of functional sites in composite fiber membranes. There are three ways to deal with them to avoid a similar situation: (1) disperse NPs uniformly in the working solution by physical methods such as stirring; (2) separately dissolve NPs and polymers in different solvents; (3) modify them using surfactants to promote uniform dispersion. With different natures of nanoparticles and polymer solutions in the mixing process, there is a large interfacial force, and the mixture can be added to a quantitative amount of surfactant, promoting the orderly arrangement of nanoparticles no longer agglomeration. Karagoz and co-workers took this approach by adding a quantitative amount of Triton x-100 (surfactant) to a mixture of ZnO nanorods and DMF, sonicated for 20 min, then added the polymer under rapid stirring until complete dissolution. Fiber diameters were uniformly distributed, and no beads were observed [81].

2.3. Formation of NNHs in the Single-Step Process

With a precursor working solution using suitable solvent and proper dispersion, the NPs are uniformly dispersed in the working solution, which requires only a single-step process to form NNHs by adjusting the ES process parameters (Figure 4A). Lopresti et al. incorporated silica (AS) or clay (CLO) NPs into 10% PLA solution and gathered them by a cylindrical grounded drum at 15 KV. The phenomenon of particle aggregation or clustering was evident with the increase in NPs content of the composite fibers (Figure 4E). Moreover, the composites with a certain number of added NPs exhibited brittleness, but bone cells diffused faster on the composites with higher particle content [82]. Abdelaziz and co-workers encapsulated silver nanoparticles (AgNPs) and hydroxyapatite (HANPs) in PCL/CA solution. Interestingly, the tensile stress of the composite fibers loaded with

10% HANPs was up to 3.39 MPa, well above that of pure PCL/CA nanofibers, but then the tensile stress of the composite fibers with 20% HANPs dropped sharply to 2.22 MPa, perhaps because the mechanical property change of NNHs with the addition of nanoparticles is a parabolic. This interesting phenomenon could contribute to the reference of optimizing the relationship between NNHs proportioning and mechanical properties [83].

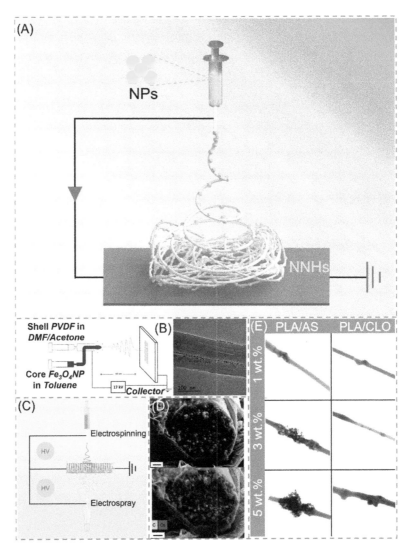

Figure 4. (**A**) Simple preparation of NNHs based on the single-step process. (**B**) Schematic diagram of coaxial preparation of NNHs and Transmission Electron Microscopy (TEM), reprinted with permission from Ref. [84]. Copyright 2021 John Wiley and Sons. (**C**) Schematic diagram of simultaneous electrospinning and electrospraying. (**D**) Scanning Electron Microscopy (SEM) images of CDP-PVP-PANI fiber cross-sections and Energy dispersive X-Ray spectroscopy map. Reprinted from Ref. [85]. (**E**) Scanning Transmission Electron Microscopy (STEM) images of PVA/AS or PVA/CLO fibers with different particle concentrations, Reprinted from Ref. [82].

The polymer solution and NPs are divided into different syringes to eliminate the agglomeration of NPs inside the fibers, which are encapsulated in the fibers by coaxial electrospinning or electrospray. Navarro Oliva and co-workers prepared core-shell-structured NNHs by the coaxial electrostatic spinning technique using an Fe_3O_4 solution as the core layer and a PVDF polymer solution as the shell layer. As shown in Figure 4B, TEM images showed that the NPs were uniformly dispersed inside the fibers without agglomeration [84]. Combining both electrospinning and electrospraying to prepare NNHs is also a convenient method. These two processes are carried out simultaneously, and NPs can be uniformly distributed in the fiber layer (Figure 4C).

The electrospinning process (ESP) has evolved from single-fluid to multi-fluid as research has progressed [86]. Yu's research team successfully prepared nanofibers with obvious core-shell structures and smoother fiber surfaces by modified tri-axial ESP and achieved intelligent three-stage controlled drug release [87,88]. Pursuing a simple and convenient preparation method, the main preparation method of NNHs is currently the uniaxial ES process. Integrating the preparation of NNHs with more sophisticated ES techniques is a challenge. The incorporation of NPs somewhat reduces the capability of NFs. There are two obvious drawbacks: (1) The addition the NPs reduces the interactions between polymer chains and decreases the mechanical properties of NNHs. (2) NPs are wrapped in fibers and cannot be functionalized. However, Radacsi et al. avoid the defect that NPs are encapsulated by nanofibers and cannot realize the specific surface area enhancement by using cesium dihydrogen phosphate (CDP) for spontaneous nucleation along the solute growth in porous nanofibers; the moist water in the air acts as a solute transport fast channel. Thus, NPs are not only grown inside the fibers but also are uniformly dispersed on the fiber surface. This strategy perfectly avoids the defect that NPs are encapsulated by nanofibers and cannot realize the specific surface area enhancement (Figure 4D) [85].

2.4. The Nanoparticles from Nanofibers through Molecular Self-Assembly

NPs are formed spontaneously by self-assembly strategies [89,90], combining this scheme with ESP, based on post-processing to form NPs spontaneously in fibers, such as in situ synthesis, hydrothermal-assistance and calcination.

2.4.1. In Situ Synthesis

The smooth surface of electrospun fibers and certain stability of the polymer, NPs cannot be perfectly and uniformly attached to the fiber surface by electrostatic adsorption or functional group action. The fiber impregnated with a ligand solution instigates bonding using a reducing agent or other reduction in the composite material to form NPs [91,92]. Lv et al. cross-linked potato starch as a polymer after forming starch fiber mats immersed in $AgNO_3$ solution, and Ag^+ was reduced to AgNPs by heating at 60 °C protected from light. The average particle size of AgNPs increased with the concentration of $AgNO_3$, but the particle size decreased significantly and agglomerated together when the concentration reached 100 mg/mL [93]. The technology is also applied to MOF-based nanofibers. Li and co-workers synthesized PW12@UiO-66 crystals in situ on PAA-PVA nanofibers. The surface of the fibers was completely covered by crystals at 12 min of growth, and the MOF crystals were enriched on the surface of NNH but with the disadvantage of random arrangement [94]. Lee and co-workers adopted a hydrodistillation-induced phase separation method to form dense cavities on the surface of PLA nanofibers, followed by uniform growth of ZIF-8 crystals on the porous fiber surface (Figure 5A [95]). Using COF materials with similar properties to MOF, Ma et al. adopted soaking PAN membranes in COF material precursor solutions, where Schiff base condensation occurred at room temperature to form COF spherical particles, and PAN@COF composite fibers showed excellent thermodynamic stability at 300 °C (Figure 5B) [91]. The NNHs prepared by this method, with dense particle aggregation, provide a sufficient number of sites of action that may provide a good solution for the adsorption of tissue fluids or other substances. Similar results from in situ synthesis can be found in Figure 5C [96] and Figure 5D [97].

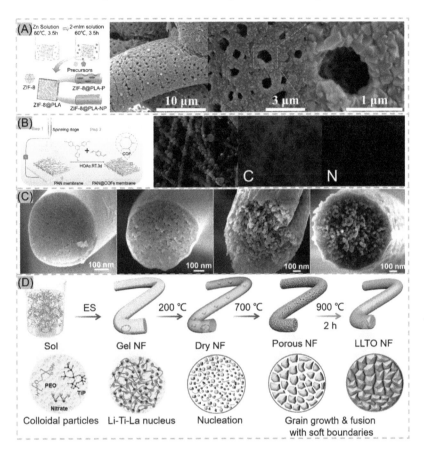

Figure 5. (**A**) Schematic of in situ ZIF-8 growth on PLA fibers and Scanning Electron Microscopy (STEM) images of ZIF@PLA-P, reprinted with permission from Ref. [95]. Copyright 2021 ACS Publications. (**B**) Schematic of PAN@COF synthesis and EDS mapping images reprinted with permission from Ref. [91]. Copyright 2022 Elsevier. (**C**) Transmission Electron Microscopy images of MBNM films prepared based on different contents of PAA, reprinted with permission from Ref. [96]. Copyright 2021 Elsevier. (**D**) Schematic of LLTO NF process prepared by electrospinning followed by calcination, reprinted with permission from Ref. [97]. Copyright 2021 John Wiley and Sons.

2.4.2. Calcination

Calcination is another effective method to prepare NNHs. Precursor fibers are prepared by the sol–gel method in combination with ESP, and calcination forms nanoparticles in the fibers [98]. Xie fabricated PVP composite fibers, which were maintained at 800 °C for 30 min, and iron-cobalt (FeCo) alloy nanoparticles (30–50 nm) were uniformly distributed in CNFs (150–300 nm), which achieved electrocatalytic degradation of antibiotics in wastewater [99]. Ding's research team proposed the use of ball-milling precursor sols to form homogeneous nuclear to precisely control crystal nucleation and growth for the purpose of grain refinement to avoid problems, such as the appearance of impurity phases and crack expansion during the preparation of flexible chalcogenide LLTO nanofibers (Figure 5D). Meanwhile, the soft grain boundary is constructed by adopting 200–900 °C stage calcination, which shows the excellent mechanical properties of flexible electronic fiber films based on perovskite ceramic oxide [92]. Nanofibers obtained by unusual calcination are brittle and inflexible. To reduce such a situation, Shan et al. formed porous structures using the template method with sacrificial PAA heated at 700 °C for 2 h to modulate the PAA concentration driving phase separation during calcination stage pyrolysis (Figure 5C) [91]. Similarly, Li and colleagues inherited the MOF structure in ZIF/PAN-Ni-15 composite nanofibers at the sacrifice of ZIFNPs by calcining at 700 °C for 2 h, followed by annealing to form a rosary morphology in the fibers [74].

2.4.3. Hydrothermal-Assistance

Hydrothermal-assistance is a common method to prepare a variety of inorganic oxide crystals, which is based on the conditions of low temperature and isobaric pressure. The uniform distribution of substances avoids the appearance of impurity phases by means of an aqueous medium, obtaining nanoparticles (such as spheres, cubes, flowers, etc.) with diverse morphologies [100–102]. The combination of hydrothermal and ESP for the preparation of multilayer heterostructures provides superior performance at a low cost and is green, easy to operate and well received by researchers. Küçük et al. immersed TiO_2 fibers (100–200 nm) in an alkaline Ba (OH)2·8H$_2$O solution by a simple hydrothermal reaction, which unexpectedly transformed the composite fibers into tetragonal crystals, demonstrating that metal oxide nanofibers can also be precursors for the preparation of $BaTiO_3$ crystals [103]. Mukhiya and colleagues grew MOF materials on PAN-prepared carbon nanomats (350 nm). First growth of ZIF-67 crystals was attributed to the hydrothermal reaction of cobalt carbonate hydroxide with 2-methylimidazole. The synthesis of ligands by deprotonated 2-methylimidazole with Co^{2+} was responsible for the crystals' second growth. The composite fiber membrane has high specific capacity and excellent service life, with strong advantages in energy storage applications [104]. In terms of the same application, Poudel et al. prepared PAN/PMMA nanofiber membranes using a coaxial ESP and sacrificed the internal PMMA after carbonization to obtain 3D hollow carbon nanofibers (3DHPCNF). By adopting two hydrothermal methods, Fe_2O_3 was first synthesized on the fiber surface, and then this was used as a precursor to hydrothermally generate ZMALDH@Fe$_2$O$_3$/3DHPCNF with ternary metal salts in an autoclave. The LDH lamellar structure was thinned by changing the content of Zn, and the transformation of nanosheets to nanowires was found at high Zn^{2+} content, which is a novel top-down way to obtain ternary LDH-electrospun hollow carbon nanofibers, providing a new idea for subsequent design [105].

3. The Biomedical Applications of Nanofiber-Nanoparticle Hybrids (NNHs)

Electrospun fibers, which are based on biomaterials, have been developed in biomedical fields such as tissue scaffolds for a long time, but the single role is not enough for sophisticated practical applications. Researchers have combined multi-functionalized NPs with NFs to develop innovative materials for drug delivery, antibacterial and tissue engineering.

3.1. Drug Delivery

Drug delivery systems (DDSs) release drugs through specific control devices and certain doses to achieve the purpose of enhancing the immunity of the body or treating diseases [106,107]. It is difficult to advance medicine without the use of medications. While drug activity is crucial, the method of administration is more so. Pharmacokinetics (PK), duration of action, metabolism and toxicity are the key elements that influence drug delivery [108]. NNHs inherit the excellent biocompatibility and high specific surface area of electrospun fibers but also the function of NPs, which can load different drugs into NNHs for a precise controlled release [109,110]. The advantages of NPs are their high specific surface area, superior bioavailability and functionalization. Similarly, in practical drug delivery applications, nanofibers can be electrospun with biocompatible polymers, resulting in minimal harmful effects on humans [111]. Tuğcu-Demiröz and co-workers employ ionic gelation to load benzydamine onto chitosan NPs. Composite NPs were embedded in NFs and hydrogels, respectively. On the one hand, NFs have a large specific surface area. The medicine, on the other hand, diffuses across a short distance in the fiber. Benzydamine released 53.03% in the composite fiber system after 24 h. Because of the strong polymer matrix, the hydrogel system only released 15.09% of benzydamine. This is insufficient for the required drug concentration for vaginal infections [112]. In the process of release, benzydamine enters the nanofiber after the initial release of chitosan nanoparticles, and the secondary release can achieve a slow release and prolong the action time.

For the mechanism of release of NNHs-loaded drugs: (1) one drug in two phases: the drug is loaded on nanoparticles that need to cross the barrier between nanoparticles and nanofibers during the release to achieve the effect of slow release. (2) Dual drug biphasic: two drugs are loaded in separate materials with different properties to achieve the required effect by different types of release mechanisms.

3.1.1. One-Drug Biphasic

One-drug biphasic is an advanced therapeutic approach to chronic diseases by precisely designing the carrier to act rapidly in the first burst and then to work longer with chronic illness. For example, He and co-workers used ethylcellulose (EC) and polyethylene glycol (PEG) small-molecule solutions in the core sheath to simultaneously load ibuprofen (IBU) through an improved coaxial ESP. The prepared engineered spindles-on-a-string (SOS) nanofiber hybrid possesses typical controlled drug release properties. When the SOS structural mixture contacts water, the hydrophilic small molecule PEG rapidly dissolves to release IBU, while the hydrophobic EC drug on the other side is continuously released in a slow-release form (Figure 6A) [113]. This precise controlled release means providing the patient with the desired drug environment in the first instance, followed by a prolonged slow release to provide effective blood levels [114–117].

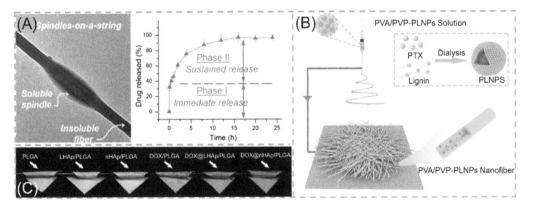

Figure 6. (**A**) Transmission electron microscopy images of engineered spindles-on-a-string (SOS) and in vitro drug dissolution curves reprinted with permission from Ref. [113]. Copyright 2021 Springer Nature. (**B**) Preparation process of PVA/PVP-PLNPs nanofiber membrane reprinted from Ref. [114]. (**C**) Digital photographs of various fiber scaffolds reprinted with permission from Ref. [116]. Copyright 2019 Elsevier.

Facing the complex living system of organisms, hydrophobic drugs confront barriers in the delivery pathway. The use of NPs as carriers can improve the solubility of hydrophobic drugs. Simultaneously, the Enhanced Permeability and Retention (EPR) effect of NPs can stabilize tumor aggregation [115]. These benefits steadily improve the use and efficacy of hydrophobic drugs [117]. Recently, Xu et al. prepared CUR/CUR@MSNs-NFs by ESP by loading curcumin (CUR) in SiO_2NPs. The scaffolds exhibited excellent biphasic release, which released 54% in the first 3 days and all in 35 days [118]. In addition, by modifying the SiO_2NPs, the amine-functionalized MSNs are hydrophilic and positively charged, resulting in a more prominent anti-cancer effect [119]. Li and co-workers used amphiphilic lignin nanoparticles (LNPs) loaded with the anticancer drug paclitaxel (PLNPs) encapsulated in PVA/PVP nanofibers, which exhibited a 59% initial abrupt release compared to the rapid abrupt release of PLNPs (Figure 6B) [115]. In the same manner, adriamycin DOX was inserted into layered nanohydroxyapatite LHAp (DOX@LHAp) to blend it with PLGA. For obtaining nanofibers, DOX was loaded with 2D layered material. The release showed significant attenuation at the initial stage of release and subsequent strong prolonged release. The in vitro release showed excellent controlled release ability (Figure 6C) [116]. However, the relevant tests are still in the laboratory stage, and the in vivo situation is more complex because there are many influencing factors. Therefore, the in vivo drug release situation needs further research.

3.1.2. Dual-Drug Biphasic Approach

Inside the face of this more complicated therapeutic environment, the role of a single drug is considerably limited. To treat this shortcoming, the dual-drug biphasic approach is adopted. The ES process is used to load various drugs onto different materials. In a dual-drug biphasic release, core-shell and Janus composite fibers are frequently employed [120]. The existence of beads on a string in the ESP is frequently unloved, with most people believing that smooth fibers and uniform particles are the desired result. Nevertheless, according to the Janus beaded fibers prepared by Li and co-workers through a homemade eccentric spinning head with a hydrophilic polymer PVP containing the MB model drug and a hydrophobic polymer EC containing ketoprofen (KET) on other side, the beads on a string have more advantages in the drug release phase compared to the prepared Janus nanofibers (Figure 7). The drug is evenly enclosed in the fibers throughout the stretching process, forming a good compartment. Because of the low viscosity of the solution during the stretching phase, the beaded structure generates an unstable jet during electrospinning. This phenomenon causes polymer aggregation to generate discrete bumps. The hydrophilic side of Janus beads releases MB quickly, whereas the hydrophobic side takes longer to release ketoprofen (IPU). This facilitates a dual-drug controlled release [121]. Then, encapsulating a nanoparticle-loaded drug in a nanofiber loaded with another drug may yield pleasantly exciting results in a controlled drug release through the difference of their properties [116]. Gupta et al. formed core-shell fibers by loading kaempferol in biodegradable albumin nanoparticles that were monolithically encapsulated in PCL loaded with dexamethasone [122]. After a weak burst release at 24 h, a sustained slow release of 15 days is possible, allowing the prolonged simultaneous action of both drugs to synergistically promote osteogenesis. Furthermore, single encapsulation of kaempferol-loaded albumin nanoparticles in PCL nanofibers with in vitro dissolution of the drug revealed a decrease in drug release rate from both composite fibers, but this is a gratifying result, where the presence of nanoparticles reduces the continuity of the fibers, resulting in a decrease in the drug release rate.

3.1.3. Smart Response Drug Delivery

Characteristics, such as site-specific rather than systemic action, controlled drug release and intelligent responsive release, need to be satisfied for effective drug delivery systems [123–125]. The advantages of NNHs are evident in blending functionalized nanoparticles within electrospun fibers, which can facilitate the combination of material benefits to achieve the desired properties [126]. Liu et al. developed a fiber surface growth particle morphology fiber through secondary growth of an MOF material by adjusting different concentrations of ligands (2-MIM) and the reaction time. The number of fiber surface particles (ZIF-8) grown produced differences in drug release analysis. When the concentration of ligands (2-MIM) is increased, the more ZIF-8 particles grow on the fiber,

Polymers **2022**, *14*, 351

leading to a decrease in the amount of drugs released. The presence of nanoparticles significantly reduced the drug release capacity. After 72 h of dissolution experiments, 89% of APS was released at pH = 5.5, and only 40% of APS was released at pH = 7.2. The relative extremely slow rate obviously depends on the pH value (Figure 8B) [127]. Croitor et al. prepared NNHs by a single-step blending process of 10% PLA with different masses of GO stirred well and by ESP [120]. The surface of the fiber is flat and smooth, and the diameter of fiber decreases significantly after loading GO nanoparticles. GO has superior electrochemical properties and can respond rapidly under external electric field stimulation. Through an in vitro drug dissolution test, the drug release capacity at 10HZ PLA/GO/Q is about 8000 times more than that without external stimulation (Figure 8C). In the treatment process, the ability to achieve an intelligent response to the characteristics of drug release through the modulation of electrical signals alone is undoubtedly a ground-breaking innovation. However, in the drug delivery process, it is necessary to provide the appropriate release rate. Too fast a release of drugs can easily reach the peak, biological activity can become to high or a certain degree of toxicity can even be produced. How to precisely regulate the ability to release drugs through changes in electrical signals is still a challenge [128].

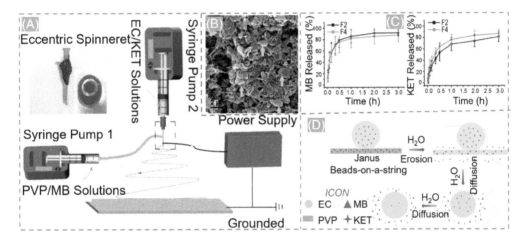

Figure 7. (**A**) Schematic diagram of Janus nanofiber preparation. (**B**) Schematic diagram of scanning electron microscopy of residual ECNPs after solubilization. (**C**) In vitro drug dissolution profiles for MB and KET. (**D**) Diagram of the drug release mechanism. Reprinted from Ref. [121].

For the same purpose, Banerjee et al. generated a PCL composite fiber containing superparamagnetic iron oxide nanoparticles (SPIONs). The NPs can carry rhodamine B model drugs. Under external stimulation, the NPs generate magnetically. Furthermore, the NPs create thermal energy as a result of Néel–Brownian relaxation [129]. The morphology of the composite under human-acceptable laser irradiation showed a molten state compared to pure PCL electrospun fibers with no significant change, which was attributed to the second effect of SPIONs (Figure 8A). In vitro assays showed that the release of rhodamine B from RF-EMF exposure showed a linear increase at a certain time, reaching 40% release after the fifth activations. Controllability is perfectly illustrated. Nevertheless, one should be aware that the magnetocaloric effect of SPIONs is complementary to thermosensitive polymers, but the generated high heat may affect the performance of the polymers as well as the biological activity of the heat-insensitive drugs [130,131].

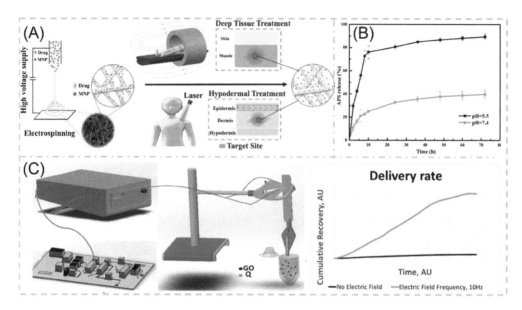

Figure 8. (A) Schematic diagram of the preparation process of magneto-thermal responsive nanofiber MSP reprinted from Ref. [129]. **(B)** Drug release profile of PH-responsive NNHs reprinted with permission from Ref. [127]. Copyright 2020 Springer Nature. **(C)** Drug release from electrically stimulated PCL/GO/Q composite nanofiber under 10 HZ electrical stimulation reprinted from Ref. [120].

3.2. Antibacterial

Recently, with the abuse of antibiotics, different types of drug-resistant bacteria emerged [39,132–134], and the combination of different types of materials based on NNHs can also generate excellent antibacterial ability against drug-resistant bacteria (Table 2). The preparation of NNHs by loading the bactericidal nanoparticles in the precursor solution begins with the mixed uniform solution, which is prepared into fibers by ESP. Mainstream bactericidal nanoparticles mainly include metal (AuNPs [135,136], AgNPs [137,138]) and metal oxides (ZnO [139], CuO [140] and TiO_2 [141]), etc. The accumulation of AgNPs in mitochondria leads to mitochondrial dysfunction. Additionally, AgNPs disrupt the DNA structure, resulting in non-replication and effectively suppressing bacterial multiplication. These two fundamental factors lead AgNPs to possess potent bactericidal potential [142]. Li et al. manufactured PCL nanofibers loaded with AgNPs and cisplatin (DDP), which can be used to prevent airway inflammation and resist granulation tissue proliferation. In comparison with the control group, the PCL-DDP-AgNPs scaffold showed no adhesion to rabbit peritracheal tissues, as shown in Figure 9D, while the coated plate had a superior antibacterial effect [137]. Yang and co-workers prepared Janus nanofibers PVP-CIP/EC-AgNPs by loading AgNPs on the EC side through bide-by-side electrospinning (Figure 9A). The Janus fibers from the disc-diffusion experiment showed a larger circle of inhibition against *S. Aureus* and *E. Coli* than fibers loaded with a single antimicrobial agent, which may be due to the simultaneous action of AgNPs and CIP, steadily increasing the antimicrobial capacity [132]. AgNPs are easily aggregated in polymer fibers, which affects the mechanical properties of the fibers and significantly compromises the antibacterial effect. Bakhsheshi-Rad et al. doped GO/AgNPs into PLLA fibers. They deposited composite fibers onto a bio-implant Mg alloy. SEM images of composite fibers co-cultured with bacteria revealed considerable bacterial cell membrane disruption (Figure 9B). When compared to PLLA nanofibers, GO/AgNPs/PLLA efficiently inhibits bacterial proliferation [133].

This strategy provides new ideas for surface modification of metallic biomaterials, but high concentrations of AgNPs possess a degree of toxicity to normal cells. There is no presence of Ag elements in the organism. For example, if they are mostly deposited inside the liver, they will cause irreversible damage to the organism. Therefore, how to effectively and precisely control the amount of AgNPs released deserves researchers' deep thinking [143]. There is a similar situation for AuNPs. Ibrahim et al. employed carboxymethyl chitosan (CMCS) as a green reducing agent for AuNPs. AuNPs were encased in PVA nanofibers. This technique reduces the cytotoxicity of AuNPs generated by the chemical reductant method while efficiently preventing bacterial growth. It signifies that safety has been improved further (Figure 9C) [135].

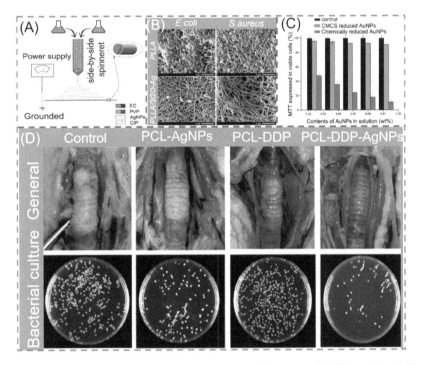

Figure 9. (**A**) Process diagram for the preparation of parallel structured NNHs, reprinted with permission from Ref. [132]. Copyright 2020 Elsevier. (**B**) Scanning electron microscopy (SEM) images of bacteria cultured on PLLA and PLLA/3GO-Ag, reprinted with permission from Ref. [133]. Copyright 2020 Elsevier. (**C**) MTT method to measure the cellular activity of various materials on epidermal cells, reprinted with permission from Ref. [135]. Copyright 2020 Elsevier. (**D**) Optical images of bronchial stents on rabbit trachea and assessment of bacterial inhibition by plate count method, reprinted from Ref. [137].

Table 2. Examples of NNHs used as antimicrobial agents.

Polymers	NPs	NNHs	Preparation Methods	Bacterial Strains	Evaluation Methodology	Antibacterial Ability	Ref.
PVA/CS	CuNPs	PVA/CS/Cu	Co-blending	S. Aureus (ATCC 25923); B. cereus (ATC 11788); E. coli (ATCC 35218); P. aeruginosa (ATCC 49189)	Antibacterial Circle	The size of the inhibition circle is: S. Aureus (15.6 ± 1.1 mm); B. cereus (29.6 ± 0.42 mm); E. coli (13.3 ± 0.8 mm); P. aeruginosa (10 ± 1 mm)	[144]
GEL/PCL/ P(DMC-AMA)	nHAP	JGM	Co-blending	E. Coli and S. Aureus	CFU Counting	The bacterial viability of S. Aureus after 6 h was 0.1%.	[145]
Starch	AgNPs	starch/AgNPs	In-situ synthesis	E. Coli (ATCC 35218); S. Aureus (ATCC 29213)	Disc Diffusion-8mm	E. Coli (9.7 mm); S. Aureus (10.2 mm)	[96]
PMMA	ZnO nanorods/ AgNPs	PMMA/ZnO-Ag NF	Co-blending, in situ synthesis	E. Coli (ATCC 25922); S. aureus (ATCC 25923)	Disc Diffusion-6mm	E. Coli (7–17 mm); S. Aureus (8.5–18.5 mm)	[81]
CH/PEO	8Ce-BG	CH-PEO-(8Ce-BG)	Co-blending	E. Coli and S. Aureus	Flat Counting Method	E. Coli activity was only 55.3%	[146]
PLLA	GO-Ag	PLLA-GO-AgNPs	Co-blending	S. Aureus (ATCC 12600); E. Coli (ATCC 9637)	Antibacterial Circle	3.01 mm–4.62 mm	[133]
PVP K90/EC	CIP/AgNPs	PVP-CIP// EC-AgNPs	Co-blending	S. Aureus (ATCC 27853); E. Coli (ATCC 25922)	Antibacterial Circle	24 h, E. Coli (17.8 ± 0.6mm mm); S. Aureus (21.9 ± 0.6 mm)	[132]
PVA	ZnO	PVA/ZnO	Self-assembly	S. Aureus (ATCC25923); E. Coli (ATCC25922)	MIC method	E. Coli (62.5 µg/mL); S. Aureus (250 µg/mL)	[147]
PLGA/SF	ZnO	PSZ	Co-blending	E. Coli and S. Aureus	turbidity measurement method	PSZ antibacterial activity against S. Aureus: 45.1–100%	[148]

3.3. Tissue Engineering

Simulating the structure and composition of the extracellular matrix (ECM) is a method to provide suitable conditions for cell adhesion, differentiation and proliferation. Researchers have found that electrospun fibers are identical in properties to natural tissues through the intersection of biological, medical and nano-engineering technologies [149,150] while offering high porosity. Electrospinning (ES) is considered one of the most eligible technologies for tissue engineering, providing scaffolds for tissues to mimic ECM composition and deliver Biofactors, thus promoting the growth of new tissues [151].

We will now discuss the functionalization of electrospun fibers by co-blending nanoparticles in the spinning system. Lu and colleagues used SiNPs deposited on the surface of PLLA electrospun fibers, which effectively enhanced the mechanical properties and hydrophilic ability of the composite membrane, leading to better biocompatibility and promoting cell adhesion and proliferation [152]. Orthopedic clinical challenges are surgical risks of bone loss aptamers and pathogenic infections. Placing autografts or an Allograft is the best clinical option to combat bone loss but with immune rejection problems and infection risks [153]. Following the development of NNHs, electrospun fiber tissue engineering scaffolds can be fabricated from nanofibers and several nanoparticle hybrids, which mainly include: inorganic nanoparticles (nanohydroxyapatite (nHA) [154], calcium phosphate (CaPs) [155], Molybdenum Disulfid (MoS$_2$) [156], Cerium dioxide (CeO$_2$) [63] and Magnesium oxide (MgO) [157]) and metal nanoparticles (gold nanoparticles (GNPs) [158]), etc.

Nanohydroxyapatite (nHA) enhances the osseointegration and osteoconductive properties of hard tissues by embedding itself in the collagen matrix of hard tissues. Song and co-workers spun NELL-1 functionalized chitosan nanoparticles (NNPs) and nHA in PCL fibers (Figure 10A). By incorporating the NPs into PCL electrospun fibers, the authors clearly increased the fiber diameter and demonstrated that nHA and NNPs enhanced cell adhesion sites. nHA presence compared to unmodified nanofibers MC3T3-E1 cell proliferation and differentiation were enhanced [159]. Liu and colleagues prepared hydrogel fibrous scaffolds containing gelatin-methacryloyl (GelMA) and calcium phosphate nano-particles (CaPs). After 14 days of incubation in SBF, significant mineralized nodules developed on the surface of CaPs@GelMA-F. Faster Ca^{2+} deposition in the early stage can significantly boost the development of calcified nodules later on (Figure 10B). Moreover, bone defects are accompanied by rupture of blood vessels. CaPs@GelMA-F cells were co-cultured with HUVECs cells, resulting in pictures of dense vascular network topology. CaPs@GelMA-F is also very effective in promoting angiogenesis [160].

Because of its unique physicochemical properties, where S is an important component of many amino acids in living organisms, MoS$_2$ is also an essential trace element in the human body [141]. Researchers found that doping MoS$_2$, a two-dimensional material, with nanofibers can effectively promote cell proliferation and maintain cell viability of BMSCs, and at the same time, the higher the MoS$_2$ content, the stronger the osteogenic ability [161,162]. Furthermore, Ma et al. found that the NIR photothermal properties possessed by MoS$_2$ can respond quickly, reaching the optimal temperature required for osteogenesis (40.5 ± 0.5 °C) in 30 s with 808 nm NIR irradiation, as shown in Figure 10C. The BV/TV capability under NIR irradiation reached 41.41 ± 0.52% [156].

In addition to the various applications in osteogenic tissues, NNHs also have a positive role for other tissue engineering formation. Neural tissue engineering (NTE) is an emerging field. Material preparation requires excellent biocompatibility, certain mechanical properties and good permeability to oxygen and nutrients, etc. [163]. Chen and co-workers demonstrated the combination of melatonin MLT and Fe$_3$O$_4$-MNP with PCL to prepare an outer-middle-internal triple-layer structural scaffold of PCL, Fe$_3$O$_4$-MNPs/PC and MLT/PCL, with MLT effectively reducing oxidative loss. The results revealed that the scaffold had a medullary axon diameter (3.30 μm) similar to the autograft group, which significantly promoted neuronal axon growth (Figure 11D) [164]. Myocardial infarction is one of the more common diseases in humans. Myocardial cells are gradually replaced by scar tissue after necrosis, leading to heart failure and arrhythmias caused by changes in electrophysiological properties, so it is most important for cardiac tissue engineering to have excellent elasticity, high electrical conductivity and promote the growth of myocardial cells [165]. Zhao et al. employed the ES process to create carbon nanotube/silk protein (CNT/silk) scaffolds. The majority of carbon nanotubes are contained within the fibers and dispersed along the fiber development direction. Compared to monospun silk protein, the elongation at the break of the composite fiber is up to 200% or more, and the tensile strength reaches 5.0 Mpa (Figure 11C). The CNT/silk stent's excellent electrical conductivity implies that it is efficient in avoiding arrhythmia (Figure 11B) [166].

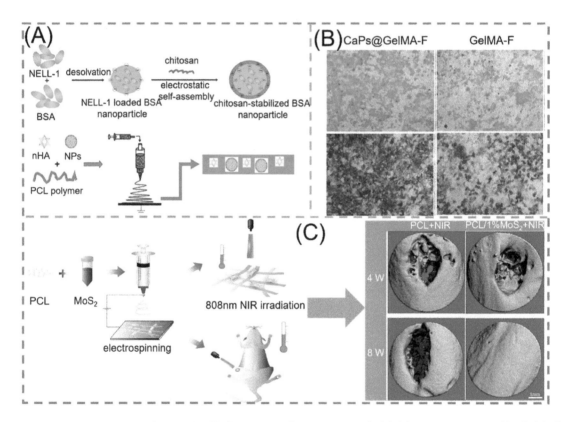

Figure 10. NNHs for promoting bone tissue growth, (**A**) Schematic preparation of loaded dual nanoparticle electrospun scaffolds, reprinted from Ref. [159]. (**B**) Osteogenic ability of loaded CaPs with unloaded electrospun fibers (after alizarin red staining and), reprinted with permission from Ref. [160]. Copyright 2020 Elsevier. (**C**) PCL and MoS$_2$ co-blended preparation of NNHs. Schematic diagram and micro-CT imaging with and without MoS$_2$ under photothermal conditions, Reprinted with permission from Ref. [156]. Copyright 2021 John Wiley and Sons.

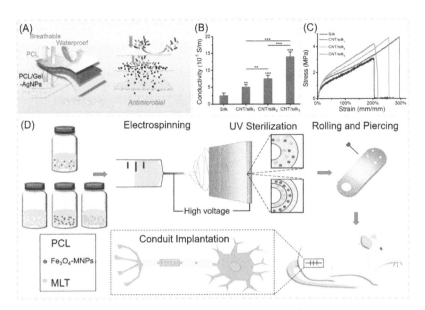

Figure 11. NNHs for other tissue engineering applications. (**A**) Schematic diagram for skin tissue engineering, Reprinted with permission from Ref. [167]. Copyright 2021 Elsevier. Tensile curves of CNT/silk scaffolds (**B**) for cardiac tissue engineering and (**C**) electrical conductivity, Reprinted with permission from Ref. [164]. Copyright 2020 John Wiley and Sons. (**D**) Process flow diagram for neural tissue engineering preparation, Reprinted with permission from ref. [164]. Copyright 2020 John Wiley and Sons.

The human immune system's first line of defense is the skin. Skin injury is common in everyday life. Wound dressings to meet the demands of skin tissue engineering are scarce. A great wound dressing must not only support ECM regeneration but also protect the skin from exogenous microorganisms [168]. NNHs fulfill the need for skin tissue engineering owing to their versatility. Janus nanofibers were loaded with AgNPS and RCSPs to impart antibacterial ability and bioactivity to electrospun fibers. In the present work, interestingly, Janus nanofibers can be obtained based on the uniaxial electrostatic spinning process through the presence of phase separation of PCL and PVP in a solution. AgNPS in the free state is very susceptible to phagocytosis by normal cells due to its fine particles and exists a certain degree of cytotoxicity, and RCSPs-Ag nanofibers in this study had only 78.86% viability on NIH 3T3 cells [169]. The team designed a sandwich wound dressing with AgNPs loaded with an intermediate layer of nanofibers to both impede the invasion of exogenous microorganisms and avoid cytotoxicity (Figure 11A) [167].

4. The Present Challenges of NNHs in Medical Applications

Nanomedicine technology is constantly evolving. The use of functionalized nanofibers and nanoparticle hybrids (NNHs) in medicine has shown encouraging effects. However, there are a number of challenges that must be surmounted: (1) Most nanoparticles and electrospun fibers necessitate the use of organic solvents in their fabrication. Organic solvent residues make the composites less biocompatible. The green chemistry of preparation should be considered. (2) The large-scale preparation of NNHs remains a challenge in the face of industrialization needs. Multi-needle collaborative electrospinning has issues with jet interactions, needle clogging and cleaning difficulties. Needle-free electrospinning method initially allows the production of large quantities of nanofibers, but they are exceedingly inhomogeneously dispersed. (3) Nanoparticles are sporadically arranged in nanofibers, and NNHs have weak mechanical properties. This makes them more fragile as tissue scaffolds for tissue engineering applications. It is vital to investigate the "bridge" that improves the interaction of NPs with the substrate. (4) Precisely controlling nanoparticle alignment in fibers is a huge difficulty. This has the potential to have a significant impact on the electron transport efficiency of NNHs in electrically sensitive drug release. This has a detrimental impact on the biosensor as well. (5) The studies on NNHs in drug release were carried out in a more ideal environment. When confronted with the organism's complicated physiological environment, more research is needed to determine how to precisely control the drug's initial burst release and duration of activity.

5. Conclusions and Outlook

In summary, combining current advanced technologies, nanofibers and nanoparticle hybrids (NNHs) have a wide range of applications in the medical field, especially in drug release, antimicrobial and tissue engineering. The electrospinning process (ESP) developed NNHs provides a convenient way to load nanoparticles (NPs) in nanofibers (NFs) in a single-step process; however, both dispersion of NPs in a polymer solution and interfacial interactions directly affect the structure and properties of NNHs, which can be sprayed directly on fibers by physical techniques (magnetron sputtering, plasma, electrospray, etc.) or chemical synthesis based on nanoparticle methods (in situ synthesis, hydrothermal-assisted, calcination, etc.) to grow NPs inside NFs.

Combining novel NPs has been attempted. Hypercrosslinked polymers with NFs are still in a preliminary stage, which is based on Friedel–Crafts reaction [170,171]. The one-step synthesis of NPs possesses controlled morphology [172], a specific surface area as high as 4000 m^2/g, excellent biocompatibility and high porosity often used for drug delivery [173]. Combining both in the medical field is beneficial for the development of new materials.

Clinical applications frequently demand scaffolds to be multifunctional, not only to enhance cell reproduction but also to suit specific needs [174]. Scaffolds for tissue engineering must also have regenerative properties, such as the capacity to promote blood vessel formation or heal torn tissues [175]. More functional NPs will be placed into NFs in the future to create novel biomaterials based on newer ES technology iterations that combine more functional particles with multi-fluid technology (three-fluid or even four-fluid ESP). The materials' synergy may result in a new form of superfunctional composite. In the future, humans will have to deal with complicated illness challenges. Personalized custom brackets must be carefully considered. The mechanical characteristics of NNHs have remained a source of contention. Three-dimensional printing technology is being used with ESP to create an artificial intervertebral disc scaffold [176]. The mechanical properties of the stent were all above the normal human threshold. Any use of NNHs in conjunction with 3D printing technology is a potential trend. Materials are highly demanding for use in living organisms. The high compression modulus of the human bionic scaffold can readily injure neighboring tissues. While the modulus is too low, it is impossible to play the bracket role [52]. Computational simulations are used to determine the best solution, and as a result, a new sector of material design for NNHs will emerge [177]. Needle-free ESP offers a novel approach to nanofiber preparation on a large scale [178,179]. The needle-free ESP used to prepare NNHs on a large scale is expected to be commercialized soon.

Author Contributions: Conceptualization, D.Y.; methodology and formal analysis, M.Z. and X.X.; writing—original draft preparation, M.Z.; writing—review and editing, W.S. and Y.T.; visualization, Y.H.; supervision and project administration, D.Y. All authors have read and agreed to the published version of the manuscript.

Funding: The APC was funded by the Natural Science Foundation of Shanghai: No. 20ZR1439000.

Data Availability Statement: Not applicable.

References

1. Koh, H.K.; Geller, A.C.; VanderWeele, T.J. Deaths From COVID-19. *J. Am. Med. Assoc.* **2021**, *325*, 133–134. [CrossRef]
2. Nalbandian, A.; Sehgal, K.; Gupta, A.; Madhavan, M.V.; McGroder, C.; Stevens, J.S.; Cook, J.R.; Nordvig, A.S.; Shalev, D.; Sehrawat, T.S.; et al. Post-Acute COVID-19 Syndrome. *Nat. Med.* **2021**, *27*, 601–615. [CrossRef] [PubMed]
3. Fauci, A.S.; Merad, M.; Swaminathan, S.; Hur, S.; Topol, E.; Fitzgerald, K.; Reis e Sousa, C.; Corbett, K.S.; Bauer, S. From MRNA Sensing to Vaccines. *Immunity* **2021**, *54*, 2676–2680. [CrossRef] [PubMed]
4. Extance, A. MRNA Vaccines: Hope beneath the Hype. *Br. Med. J.* **2021**, *375*, n2744. [CrossRef]
5. El Sahly, H.M.; Baden, L.R.; Essink, B.; Doblecki-Lewis, S.; Martin, J.M.; Anderson, E.J.; Campbell, T.B.; Clark, J.; Jackson, L.A.; Fichtenbaum, C.J.; et al. Efficacy of the MRNA-1273 SARS-CoV-2 Vaccine at Completion of Blinded Phase. *N. Engl. J. Med.* **2021**, *385*, 1774–1785. [CrossRef]
6. Ali, K.; Berman, G.; Zhou, H.; Deng, W.; Faughnan, V.; Coronado-Voges, M.; Ding, B.; Dooley, J.; Girard, B.; Hillebrand, W.; et al. Evaluation of MRNA-1273 SARS-CoV-2 Vaccine in Adolescents. *N. Engl. J. Med.* **2021**, *385*, 2241–2251. [CrossRef]
7. Becker, M.L.; Burdick, J.A. Introduction: Polymeric Biomaterials. *Chem. Rev.* **2021**, *121*, 10789–10791. [CrossRef]
8. Safian, M.T.; Umar, K.; Parveen, T.; Yaqoob, A.A.; Ibrahim, M.N.M. Chapter Eight-Biomedical applications of smart polymer composites. In *Smart Polymer Nanocomposites: Biomedical and Environmental Applications*; Elsevier Inc.: Cambridge, MA, USA, 2021; pp. 183–204.
9. Chakraborty, P.; Oved, H.; Bychenko, D.; Yao, Y.; Tang, Y.; Zilberzwige-Tal, S.; Wei, G.; Dvir, T.; Gazit, E. Nanoengineered Peptide-Based Antimicrobial Conductive Supramolecular Biomaterial for Cardiac Tissue Engineering. *Adv. Mater.* **2021**, *33*, 2008715. [CrossRef]
10. Kang, S.; Hou, S.; Chen, X.; Yu, D.G.; Wang, L.; Li, X.; Williams, G.R.R. Energy-saving electrospinning with a concentric Teflon-core rod spinneret to create medicated nanofibers. *Polymers* **2020**, *12*, 2421. [CrossRef] [PubMed]
11. Yaqoob, A.A.; Safian, M.T.; Rashid, M.; Parveen, T.; Umar, K.; Ibrahim, M.N.M. Chapter One-Introduction of Smart Polymer Nanocomposites. In *Smart Polymer Nanocomposites: Biomedical and Environmental Applications*; Elsevier Inc.: Cambridge, MA, USA, 2021; pp. 1–25.
12. Jung, K.; Corrigan, N.; Wong, E.H.H.; Boyer, C. Bioactive Synthetic Polymers. *Adv. Mater.* **2021**, *34*, 2105063. [CrossRef] [PubMed]
13. Sionkowska, A. Collagen Blended with Natural Polymers: Recent Advances and Trends. *Prog. Polym. Sci.* **2021**, *122*, 101452. [CrossRef]
14. Joyce, K.; Fabra, G.T.; Bozkurt, Y.; Pandit, A. Bioactive Potential of Natural Biomaterials: Identification, Retention and Assessment of Biological Properties. *Signal Transduc. Target. Ther.* **2021**, *6*, 1–28. [CrossRef]
15. Rosén, T.; Hsiao, B.S.; Söderberg, L.D. Elucidating the Opportunities and Challenges for Nanocellulose Spinning. *Adv. Mater.* **2021**, *33*, 2001238. [CrossRef]
16. Taokaew, S.T.; Chiaoprakobkij, N.; Siripong, P.; Sanchavanakit, N.; Pavasant, P.; Phisalaphong, M. Multifunctional Cellulosic Nanofiber Film with Enhanced Antimicrobial and Anticancer Properties by Incorporation of Ethanolic Extract of Garcinia Mangostana Peel. *Mater. Sci. Eng. C* **2021**, *120*, 111783. [CrossRef]
17. Chen, Z.; Zhang, Q.; Li, H.; Wei, Q.; Zhao, X.; Chen, F. Elastin-like Polypeptide Modified Silk Fibroin Porous Scaffold Promotes Osteochondral Repair. *Bioact. Mater.* **2021**, *6*, 589–601. [CrossRef]
18. Bakhshandeh, B.; Nateghi, S.S.; Gazani, M.M.; Dehghani, Z.; Mohammadzadeh, F. A Review on Advances in the Applications of Spider Silk in Biomedical Issues. *Int. J. Biol. Macromol.* **2021**, *192*, 258–271. [CrossRef]
19. Dziadek, M.; Dziadek, K.; Checinska, K.; Zagrajczuk, B.; Golda-Cepa, M.; Brzychczy-Wloch, M.; Menaszek, E.; Kopec, A.; Cholewa-Kowalska, K. PCL and PCL/Bioactive Glass Biomaterials as Carriers for Biologically Active Polyphenolic Compounds: Comprehensive Physicochemical and Biological Evaluation. *Bioact. Mater.* **2021**, *6*, 1811–1826. [CrossRef]
20. Janmohammadi, M.; Nourbakhsh, M.S. Electrospun Polycaprolactone Scaffolds for Tissue Engineering: A Review. *Int. J. Polym. Mater. Polym. Biomater.* **2019**, *68*, 527–539. [CrossRef]
21. Zavan, B.; Gardin, C.; Guarino, V.; Rocca, T.; Cruz Maya, I.; Zanotti, F.; Ferroni, L.; Brunello, G.; Chachques, J.-C.; Ambrosio, L.; et al. Electrospun PCL-Based Vascular Grafts: In Vitro Tests. *Nanomaterials* **2021**, *11*, 751. [CrossRef]
22. Lv, H.; Guo, S.; Zhang, G.; He, W.; Wu, Y.; Yu, D.-G. Electrospun Structural Hybrids of Acyclovir-Polyacrylonitrile at Acyclovir for Modifying Drug Release. *Polymers* **2021**, *13*, 4286. [CrossRef]
23. Zhang, X.; Guo, S.; Qin, Y.; Li, C. Functional Electrospun Nanocomposites for Efficient Oxygen Reduction Reaction. *Chem. Res. Chin. Univ.* **2021**, *37*, 379–393. [CrossRef]
24. Ning, T.; Zhou, Y.; Xu, H.; Guo, S.; Wang, K.; Yu, D.-G. Orodispersible Membranes from a Modified Coaxial Electrospinning for Fast Dissolution of Diclofenac Sodium. *Membranes* **2021**, *11*, 802. [CrossRef]
25. Xu, H.; Xu, X.; Li, S.; Song, W.-L.; Yu, D.-G.; Annie Bligh, S.W. The Effect of Drug Heterogeneous Distributions within Core-Sheath Nanostructures on Its Sustained Release Profiles. *Biomolecules* **2021**, *11*, 1330. [CrossRef]
26. Wang, Y.; Xu, H.; Wu, M.; Yu, D.-G. Nanofibers-Based Food Packaging. *ES Food Agrofor.* **2022**, *2*, 1–23. [CrossRef]
27. Xue, J.; Wu, T.; Dai, Y.; Xia, Y. Electrospinning and Electrospun Nanofibers: Methods, Materials, and Applications. *Chem. Rev.* **2019**, *119*, 5298–5415. [CrossRef]
28. Liu, X.; Xu, H.; Zhang, M.; Yu, D.-G. Electrospun Medicated Nanofibers for Wound Healing: Review. *Membranes* **2021**, *11*, 770. [CrossRef]

29. Kang, S.; Zhao, K.; Yu, D.G.; Zheng, X.; Huang, C. Advances in Biosensing and Environmental Monitoring Based on Electrospun Nanofibers. *Adv. Fiber Mater.* **2022**, *9*. [CrossRef]
30. Zhao, K.; Kang, S.-X.; Yang, Y.-Y.; Yu, D.-G. Electrospun Functional Nanofiber Membrane for Antibiotic Removal in Water: Review. *Polymers* **2021**, *13*, 226. [CrossRef]
31. Shepa, I.; Mudra, E.; Dusza, J. Electrospinning through the Prism of Time. *Mater. Today Chem.* **2021**, *21*, 100543. [CrossRef]
32. Madruga, L.Y.C.; Kipper, M.J. Expanding the Repertoire of Electrospinning: New and Emerging Biopolymers, Techniques, and Applications. *Adv. Heal. Mater.* **2021**, 2101979. [CrossRef]
33. Zhang, X.; Xie, L.; Wang, X.; Shao, Z.; Kong, B. Electrospinning Super-Assembly of Ultrathin Fibers from Single- to Multi-Taylor Cone Sites. *Appl. Mater. Today* **2021**, *23*, 101272. [CrossRef]
34. Wang, X.-X.; Yu, G.-F.; Zhang, J.; Yu, M.; Ramakrishna, S.; Long, Y.-Z. Conductive Polymer Ultrafine Fibers via Electrospinning: Preparation, Physical Properties and Applications. *Prog. Mater. Sci.* **2021**, *115*, 100704. [CrossRef]
35. Filip, P.; Zelenkova, J.; Peer, P. Electrospinning of a Copolymer PVDF-Co-HFP Solved in DMF/Acetone: Explicit Relations among Viscosity, Polymer Concentration, DMF/Acetone Ratio and Mean Nanofiber Diameter. *Polymers* **2021**, *13*, 3418. [CrossRef] [PubMed]
36. Li, S.; Kong, L.; Ziegler, G.R. Electrospinning of Octenylsuccinylated Starch-Pullulan Nanofibers from Aqueous Dispersions. *Carbohydr. Polym.* **2021**, *258*, 116933. [CrossRef] [PubMed]
37. Bai, J.; Wang, S.; Li, Y.; Wang, Z.; Tang, J. Effect of Chemical Structure and Molecular Weight on the Properties of Lignin-Based Ultrafine Carbon Fibers. *Int. J. Biol. Macromol.* **2021**, *187*, 594–602. [CrossRef] [PubMed]
38. Fan, Y.; Tian, X.; Zheng, L.; Jin, X.; Zhang, Q.; Xu, S.; Liu, P.; Yang, N.; Bai, H.; Wang, H. Yeast Encapsulation in Nanofiber via Electrospinning: Shape Transformation, Cell Activity and Immobilized Efficiency. *Mater. Sci. Eng. C* **2021**, *120*, 111747. [CrossRef] [PubMed]
39. Ma, J.; Xu, C.; Yu, H.; Feng, Z.; Yu, W.; Gu, L.; Liu, Z.; Chen, L.; Jiang, Z.; Hou, J. Electro-Encapsulation of Probiotics in Gum Arabic-Pullulan Blend Nanofibres Using Electrospinning Technology. *Food Hydrocoll.* **2021**, *111*, 106381. [CrossRef]
40. Silva, P.M.; Prieto, C.; Lagarón, J.M.; Pastrana, L.M.; Coimbra, M.A.; Vicente, A.A.; Cerqueira, M.A. Food-Grade Hydroxypropyl Methylcellulose-Based Formulations for Electrohydrodynamic Processing: Part I–Role of Solution Parameters on Fibre and Particle Production. *Food Hydrocoll.* **2021**, *118*, 106761. [CrossRef]
41. Topuz, F.; Abdulhamid, M.A.; Holtzl, T.; Szekely, G. Nanofiber Engineering of Microporous Polyimides through Electrospinning: Influence of Electrospinning Parameters and Salt Addition. *Mater. Des.* **2021**, *198*, 109280. [CrossRef]
42. Joy, N.; Anuraj, R.; Viravalli, A.; Dixit, H.N.; Samavedi, S. Coupling between Voltage and Tip-to-Collector Distance in Polymer Electrospinning: Insights from Analysis of Regimes, Transitions and Cone/Jet Features. *Chem. Eng. Sci.* **2021**, *230*, 116200. [CrossRef]
43. Lei, L.; Chen, S.; Nachtigal, C.J.; Moy, T.F.; Yong, X.; Singer, J.P. Homogeneous Gelation Leads to Nanowire Forests in the Transition between Electrospray and Electrospinning. *Mater. Horiz.* **2020**, *7*, 2643–2650. [CrossRef]
44. Lv, S.; Zhang, Y.; Jiang, L.; Zhao, L.; Wang, J.; Liu, F.; Wang, C.; Yan, X.; Sun, P.; Wang, L.; et al. Mixed Potential Type YSZ-Based NO_2 Sensors with Efficient Three-Dimensional Three-Phase Boundary Processed by Electrospinning. *Sens. Actuators B Chem.* **2022**, *354*, 131219. [CrossRef]
45. Wang, C.; Zuo, Q.; Wang, L.; Long, B.; Salleh, K.M.; Anuar, N.I.S.; Zakaria, S. Diameter Optimization of Polyvinyl Alcohol/Sodium Alginate Fiber Membranes Using Response Surface Methodology. *Mater. Chem. Phys.* **2021**, *271*, 124969. [CrossRef]
46. Kopp, A.; Smeets, R.; Gosau, M.; Kröger, N.; Fuest, S.; Köpf, M.; Kruse, M.; Krieger, J.; Rutkowski, R.; Henningsen, A.; et al. Effect of Process Parameters on Additive-Free Electrospinning of Regenerated Silk Fibroin Nonwovens. *Bioact. Mater.* **2020**, *5*, 241–252. [CrossRef]
47. Wu, L.; Zhang, Y.; Yu, H.; Jia, Y.; Wang, X.; Ding, B. Self-Assembly of Polyethylene Oxide and Its Composite Nanofibrous Membranes with Cellular Network Structure. *Compos. Commun.* **2021**, *27*, 100759. [CrossRef]
48. Wang, Y.; Chen, G.; Zhang, H.; Zhao, C.; Sun, L.; Zhao, Y. Emerging Functional Biomaterials as Medical Patches. *ACS Nano* **2021**, *15*, 5977–6007. [CrossRef]
49. Huo, P.; Han, X.; Zhang, W.; Zhang, J.; Kumar, P.; Liu, B. Electrospun Nanofibers of Polycaprolactone/Collagen as a Sustained-Release Drug Delivery System for Artemisinin. *Pharmaceutics* **2021**, *13*, 1228. [CrossRef]
50. Chen, K.; Pan, H.; Yan, Z.; Li, Y.; Ji, D.; Yun, K.; Su, Y.; Liu, D.; Pan, W. A Novel Alginate/Gelatin Sponge Combined with Curcumin-Loaded Electrospun Fibers for Postoperative Rapid Hemostasis and Prevention of Tumor Recurrence. *Int. J. Biol. Macromol.* **2021**, *182*, 1339–1350. [CrossRef]
51. Zhang, X.; Li, L.; Ouyang, J.; Zhang, L.; Xue, J.; Zhang, H.; Tao, W. Electroactive Electrospun Nanofibers for Tissue Engineering. *Nano Today* **2021**, *39*, 101196. [CrossRef]
52. Zhu, M.; Tan, J.; Liu, L.; Tian, J.; Li, L.; Luo, B.; Zhou, C.; Lu, L. Construction of Biomimetic Artificial Intervertebral Disc Scaffold via 3D Printing and Electrospinning. *Mater. Sci. Eng. C* **2021**, *128*, 112310. [CrossRef]
53. Rickel, A.P.; Deng, X.; Engebretson, D.; Hong, Z. Electrospun Nanofiber Scaffold for Vascular Tissue Engineering. *Mater. Sci. Eng. C* **2021**, *129*, 112373. [CrossRef]
54. Wang, Z.; Li, J.; Shao, C.; Lin, X.; Yang, Y.; Chen, N.; Wang, Y.; Qu, L. Moisture Power in Natural Polymeric Silk Fibroin Flexible Membrane Triggers Efficient Antibacterial Activity of Silver Nanoparticles. *Nano Energy* **2021**, *90*, 106529. [CrossRef]

55. Ho Na, J.; Chan Kang, Y.; Park, S.-K. Electrospun MOF-Based ZnSe Nanocrystals Confined in N-Doped Mesoporous Carbon Fibers as Anode Materials for Potassium Ion Batteries with Long-Term Cycling Stability. *Chem. Eng. J.* **2021**, *425*, 131651. [CrossRef]

56. Zhao, K.; Lu, Z.-H.; Zhao, P.; Kang, S.-X.; Yang, Y.-Y.; Yu, D.-G. Modified Tri–Axial Electrospun Functional Core–Shell Nanofibrous Membranes for Natural Photodegradation of Antibiotics. *Chem. Eng. J.* **2021**, *425*, 131455. [CrossRef]

57. Huang, G.-Y.; Chang, W.-J.; Lu, T.-W.; Tsai, I.-L.; Wu, S.-J.; Ho, M.-H.; Mi, F.-L. Electrospun CuS Nanoparticles/Chitosan Nanofiber Composites for Visible and near-Infrared Light-Driven Catalytic Degradation of Antibiotic Pollutants. *Chem. Eng. J.* **2022**, *431*, 134059. [CrossRef]

58. Cui, Z.; Shen, S.; Yu, J.; Si, J.; Cai, D.; Wang, Q. Electrospun Carbon Nanofibers Functionalized with NiCo2S4 Nanoparticles as Lightweight, Flexible and Binder-Free Cathode for Aqueous Ni-Zn Batteries. *Chem. Eng. J.* **2021**, *426*, 130068. [CrossRef]

59. Sharma, D.; Satapathy, B.K. Polymer Substrate-Based Transition Metal Modified Electrospun Nanofibrous Materials: Current Trends in Functional Applications and Challenges. *Polym. Rev.* **2021**, *61*, 1–46. [CrossRef]

60. Jena, S.K.; Sadasivam, R.; Packirisamy, G.; Saravanan, P. Nanoremediation: Sunlight Mediated Dye Degradation Using Electrospun PAN/CuO–ZnO Nanofibrous Composites. *Environ. Pollut.* **2021**, *280*, 116964. [CrossRef]

61. Ding, C.; Breunig, M.; Timm, J.; Marschall, R.; Senker, J.; Agarwal, S. Flexible, Mechanically Stable, Porous Self-Standing Microfiber Network Membranes of Covalent Organic Frameworks: Preparation Method and Characterization. *Adv. Funct. Mater.* **2021**, *31*, 2106507. [CrossRef]

62. Neibolts, N.; Platnieks, O.; Gaidukovs, S.; Barkane, A.; Thakur, V.K.; Filipova, I.; Mihai, G.; Zelca, Z.; Yamaguchi, K.; Enachescu, M. Needle-Free Electrospinning of Nanofibrillated Cellulose and Graphene Nanoplatelets Based Sustainable Poly (Butylene Succinate) Nanofibers. *Mater. Today Chem.* **2020**, *17*, 100301. [CrossRef]

63. Ren, S.; Zhou, Y.; Zheng, K.; Xu, X.; Yang, J.; Wang, X.; Miao, L.; Wei, H.; Xu, Y. Cerium Oxide Nanoparticles Loaded Nanofibrous Membranes Promote Bone Regeneration for Periodontal Tissue Engineering. *Bioact. Mater.* **2022**, *7*, 242–253. [CrossRef]

64. Rasekh, A.; Raisi, A. Electrospun Nanofibrous Polyether-Block-Amide Membrane Containing Silica Nanoparticles for Water Desalination by Vacuum Membrane Distillation. *Sep. Purif. Technol.* **2021**, *275*, 119149. [CrossRef]

65. Jiang, X.; Zhang, S.; Zou, B.; Li, G.; Yang, S.; Zhao, Y.; Lian, J.; Li, H.; Ji, H. Electrospun CoSe@NC Nanofiber Membrane as an Effective Polysulfides Adsorption-Catalysis Interlayer for Li-S Batteries. *Chem. Eng. J.* **2022**, *430*, 131911. [CrossRef]

66. Fang, H.; Wang, C.; Li, D.; Zhou, S.; Du, Y.; Zhang, H.; Hang, C.; Tian, Y.; Suga, T. Fabrication of Ag@Ag2O-MnOx Composite Nanowires for High-Efficient Room-Temperature Removal of Formaldehyde. *J. Mater. Sci. Technol.* **2021**, *91*, 5–16. [CrossRef]

67. Park, K.; Kang, S.; Park, J.; Hwang, J. Fabrication of Silver Nanowire Coated Fibrous Air Filter Medium via a Two-Step Process of Electrospinning and Electrospray for Anti-Bioaerosol Treatment. *J. Hazard. Mater.* **2021**, *411*, 125043. [CrossRef]

68. Enculescu, M.; Costas, A.; Evanghelidis, A.; Enculescu, I. Fabrication of ZnO and TiO2 Nanotubes via Flexible Electro-Spun Nanofibers for Photocatalytic Applications. *Nanomaterials* **2021**, *11*, 1305. [CrossRef]

69. Wang, K.; Wang, P.; Wang, M.; Yu, D.-G.; Wan, F.; Bligh, S.W.A. Comparative Study of Electrospun Crystal-Based and Composite-Based Drug Nano Depots. *Mater. Sci. Eng. C* **2020**, *113*, 110988. [CrossRef]

70. Wang, M.; Hou, J.; Yu, D.-G.; Li, S.; Zhu, J.; Chen, Z. Electrospun Tri-Layer Nanodepots for Sustained Release of Acyclovir. *J. Alloy. Compd.* **2020**, *846*, 156471. [CrossRef]

71. Zheng, G.; Peng, H.; Jiang, J.; Kang, G.; Liu, J.; Zheng, J.; Liu, Y. Surface Functionalization of PEO Nanofibers Using a TiO2 Suspension as Sheath Fluid in a Modified Coaxial Electrospinning Process. *Chem. Res. Chin. Univ.* **2021**, *37*, 571–577. [CrossRef]

72. Yang, C.-H.; Hsiao, Y.-C.; Lin, L.-Y. Novel in Situ Synthesis of Freestanding Carbonized ZIF67/Polymer Nanofiber Electrodes for Supercapacitors via Electrospinning and Pyrolysis Techniques. *ACS Appl. Mater. Interfaces* **2021**, *13*, 41637–41648. [CrossRef]

73. Pant, B.; Prasad Ojha, G.; Acharya, J.; Park, M. Ag3PO4-TiO2-Carbon Nanofiber Composite: An Efficient Visible-Light Photocatalyst Obtained from Eelectrospinning and Hydrothermal Methods. *Sep. Purif. Technol.* **2021**, *276*, 119400. [CrossRef]

74. Li, R.; Ke, H.; Shi, C.; Long, Z.; Dai, Z.; Qiao, H.; Wang, K. Mesoporous RGO/NiCo2O4@carbon Composite Nanofibers Derived from Metal-Organic Framework Compounds for Lithium Storage. *Chem. Eng. J.* **2021**, *415*, 128874. [CrossRef]

75. Alturki, A.M. Rationally Design of Electrospun Polysaccharides Polymeric Nanofiber Webs by Various Tools for Biomedical Applications: A Review. *Int. J. Biol. Macromol.* **2021**, *184*, 648–665. [CrossRef] [PubMed]

76. Ding, J.; Zhang, J.; Li, J.; Li, D.; Xiao, C.; Xiao, H.; Yang, H.; Zhuang, X.; Chen, X. Electrospun Polymer Biomaterials. *Prog. Polym. Sci.* **2019**, *90*, 1–34. [CrossRef]

77. Valerini, D.; Tammaro, L.; Vitali, R.; Guillot, G.; Rinaldi, A. Sputter-Deposited Ag Nanoparticles on Electrospun PCL Scaffolds: Morphology, Wettability and Antibacterial Activity. *Coatings* **2021**, *11*, 345. [CrossRef]

78. Liu, Z.; Jia, L.; Yan, Z.; Bai, L. Plasma-Treated Electrospun Nanofibers as a Template for the Electrostatic Assembly of Silver Nanoparticles. *N. J. Chem.* **2018**, *42*, 11185–11191. [CrossRef]

79. Rostamabadi, H.; Falsafi, S.R.; Rostamabadi, M.M.; Assadpour, E.; Jafari, S.M. Electrospraying as a Novel Process for the Synthesis of Particles/Nanoparticles Loaded with Poorly Water-Soluble Bioactive Molecules. *Adv. Colloid Interface Sci.* **2021**, *290*, 102384. [CrossRef]

80. Fahimirad, S.; Abtahi, H.; Satei, P.; Ghaznavi-Rad, E.; Moslehi, M.; Ganji, A. Wound Healing Performance of PCL/Chitosan Based Electrospun Nanofiber Electrosprayed with Curcumin Loaded Chitosan Nanoparticles. *Carbohydr. Polym.* **2021**, *259*, 117640. [CrossRef] [PubMed]

81. Karagoz, S.; Kiremitler, N.B.; Sarp, G.; Pekdemir, S.; Salem, S.; Goksu, A.G.; Onses, M.S.; Sozdutmaz, I.; Sahmetlioglu, E.; Ozkara, E.S.; et al. Antibacterial, Antiviral, and Self-Cleaning Mats with Sensing Capabilities Based on Electrospun Nanofibers Decorated with ZnO Nanorods and Ag Nanoparticles for Protective Clothing Applications. *ACS Appl. Mater. Interfaces* **2021**, *13*, 5678–5690. [CrossRef]
82. Lopresti, F.; Pavia, F.C.; Ceraulo, M.; Capuana, E.; Brucato, V.; Ghersi, G.; Botta, L.; La Carrubba, V. Physical and Biological Properties of Electrospun Poly(d,l-Lactide)/Nanoclay and Poly(d,l-Lactide)/Nanosilica Nanofibrous Scaffold for Bone Tissue Engineering. *J. Biomed. Mater. Res. Part A* **2021**, *109*, 2120–2136. [CrossRef]
83. Abdelaziz, D.; Hefnawy, A.; Al-Wakeel, E.; El-Fallal, A.; El-Sherbiny, I.M. New Biodegradable Nanoparticles-in-Nanofibers Based Membranes for Guided Periodontal Tissue and Bone Regeneration with Enhanced Antibacterial Activity. *J. Adv. Res.* **2021**, *28*, 51–62. [CrossRef]
84. Navarro Oliva, F.S.; Picart, L.; Leon-Valdivieso, C.Y.; Benalla, A.; Lenglet, L.; Ospina, A.; Jestin, J.; Bedoui, F. Coaxial Electrospinning Process toward Optimal Nanoparticle Dispersion in Polymeric Matrix. *Polym. Compos.* **2021**, *42*, 1565–1573. [CrossRef]
85. Radacsi, N.; Campos, F.D.; Chisholm, C.R.I.; Giapis, K.P. Spontaneous Formation of Nanoparticles on Electrospun Nanofibres. *Nat. Commun.* **2018**, *9*, 4740. [CrossRef]
86. Yu, D.G.; Lv, H. Preface-striding into Nano Drug Delivery. *Curr. Drug Deliv.* **2022**, *19*, 1–3.
87. Ding, Y.; Dou, C.; Chang, S.; Xie, Z.; Yu, D.-G.; Liu, Y.; Shao, J. Core–Shell Eudragit S100 Nanofibers Prepared via Triaxial Electrospinning to Provide a Colon-Targeted Extended Drug Release. *Polymers* **2020**, *12*, 2034. [CrossRef] [PubMed]
88. Chang, S.; Wang, M.; Zhang, F.; Liu, Y.; Liu, X.; Yu, D.-G.; Shen, H. Sheath-Separate-Core Nanocomposites Fabricated Using a Trifluid Electrospinning. *Mater. Des.* **2020**, *192*, 108782. [CrossRef]
89. Chen, J.; Pan, S.; Zhou, J.; Lin, Z.; Qu, Y.; Glab, A.; Han, Y.; Richardson, J.J.; Caruso, F. Assembly of Bioactive Nanoparticles via Metal–Phenolic Complexation. *Adv. Mater.* **2021**, *33*, 2108624. [CrossRef] [PubMed]
90. Zhang, Z.; Yue, Y.-X.; Xu, L.; Wang, Y.; Geng, W.-C.; Li, J.-J.; Kong, X.; Zhao, X.; Zheng, Y.; Zhao, Y.; et al. Macrocyclic-Amphiphile-Based Self-Assembled Nanoparticles for Ratiometric Delivery of Therapeutic Combinations to Tumors. *Adv. Mater.* **2021**, *33*, 2007719. [CrossRef] [PubMed]
91. Ma, J.; Yu, Z.; Liu, S.; Chen, Y.; Lv, Y.; Liu, Y.; Lin, C.; Ye, X.; Shi, Y.; Liu, M.; et al. Efficient Extraction of Trace Organochlorine Pesticides from Environmental Samples by a Polyacrylonitrile Electrospun Nanofiber Membrane Modified with Covalent Organic Framework. *J. Hazard. Mater.* **2022**, *424*, 127455. [CrossRef] [PubMed]
92. Meng, Q.; Huang, Y.; Ye, J.; Xia, G.; Wang, G.; Dong, L.; Yang, Z.; Yu, X. Electrospun Carbon Nanofibers with In-Situ Encapsulated Ni Nanoparticles as Catalyst for Enhanced Hydrogen Storage of MgH$_2$. *J. Alloys Compd.* **2021**, *851*, 156874. [CrossRef]
93. Lv, H.; Cui, S.; Yang, Q.; Song, X.; Wang, D.; Hu, J.; Zhou, Y.; Liu, Y. AgNPs-Incorporated Nanofiber Mats: Relationship between AgNPs Size/Content, Silver Release, Cytotoxicity, and Antibacterial Activity. *Mater. Sci. Eng. C* **2021**, *118*, 111331. [CrossRef] [PubMed]
94. Li, T.; Zhang, Z.; Liu, L.; Gao, M.; Han, Z. A Stable Metal-Organic Framework Nanofibrous Membrane as Photocatalyst for Simultaneous Removal of Methyl Orange and Formaldehyde from Aqueous Solution. *Colloids Surf. A* **2021**, *617*, 126359. [CrossRef]
95. Lee, J.; Jung, S.; Park, H.; Kim, J. Bifunctional ZIF-8 Grown Webs for Advanced Filtration of Particulate and Gaseous Matters: Effect of Charging Process on the Electrostatic Capture of Nanoparticles and Sulfur Dioxide. *ACS Appl. Mater. Interfaces* **2021**, *13*, 50401–50410. [CrossRef] [PubMed]
96. Shan, H.; Si, Y.; Yu, J.; Ding, B. Facile Access to Highly Flexible and Mesoporous Structured Silica Fibrous Membranes for Tetracyclines Removal. *Chem. Eng. J.* **2021**, *417*, 129211. [CrossRef]
97. Li, X.; Zhang, Y.; Zhang, L.; Xia, S.; Zhao, Y.; Yan, J.; Yu, J.; Ding, B. Synthesizing Superior Flexible Oxide Perovskite Ceramic Nanofibers by Precisely Controlling Crystal Nucleation and Growth. *Small* **2021**, *11*, 2106500. [CrossRef]
98. Lee, S.S.; Bai, H.; Chua, S.C.; Lee, K.W.; Sun, D.D. Electrospun Bi^{3+}/TiO$_2$ Nanofibers for Concurrent Photocatalytic H$_2$ and Clean Water Production from Glycerol under Solar Irradiation: A Systematic Study. *J. Cleaner Prod.* **2021**, *298*, 126671. [CrossRef]
99. Xie, W.; Shi, Y.; Wang, Y.; Zheng, Y.; Liu, H.; Hu, Q.; Wei, S.; Gu, H.; Guo, Z. Electrospun Iron/Cobalt Alloy Nanoparticles on Carbon Nanofibers towards Exhaustive Electrocatalytic Degradation of Tetracycline in Wastewater. *Chem. Eng. J.* **2021**, *405*, 126585. [CrossRef]
100. Chang, Y.; Dong, C.; Zhou, D.; Li, A.; Dong, W.; Cao, X.-Z.; Wang, G. Fabrication and Elastic Properties of TiO$_2$ Nanohelix Arrays through a Pressure-Induced Hydrothermal Method. *ACS Nano* **2021**, *15*, 14174–14184. [CrossRef]
101. Jung, K.-W.; Lee, S.Y.; Choi, J.-W.; Hwang, M.-J.; Shim, W.G. Synthesis of Mg–Al Layered Double Hydroxides-Functionalized Hydrochar Composite via an in Situ One-Pot Hydrothermal Method for Arsenate and Phosphate Removal: Structural Characterization and Adsorption Performance. *Chem. Eng. J.* **2021**, *420*, 129775. [CrossRef]
102. Xu, W.; Li, X.; Peng, C.; Yang, G.; Cao, Y.; Wang, H.; Peng, F.; Yu, H. One-Pot Synthesis of Ru/Nb$_2$O$_5$@Nb$_2$C Ternary Photocatalysts for Water Splitting by Harnessing Hydrothermal Redox Reactions. *Appl. Catal. B* **2022**, *303*, 120910. [CrossRef]
103. Küçük, Ö.; Teber, S.; Cihan Kaya, İ.; Akyıldız, H.; Kalem, V. Photocatalytic Activity and Dielectric Properties of Hydrothermally Derived Tetragonal BaTiO$_3$ Nanoparticles Using TiO$_2$ Nanofibers. *J. Alloys Compd.* **2018**, *765*, 82–91. [CrossRef]
104. Mukhiya, T.; Tiwari, A.P.; Chhetri, K.; Kim, T.; Dahal, B.; Muthurasu, A.; Kim, H.Y. A Metal–Organic Framework Derived Cobalt Oxide/Nitrogen-Doped Carbon Nanotube Nanotentacles on Electrospun Carbon Nanofiber for Electrochemical Energy Storage. *Chem. Eng. J.* **2021**, *420*, 129679. [CrossRef]

105. Poudel, M.B.; Kim, H.J. Confinement of Zn-Mg-Al-Layered Double Hydroxide and α-Fe$_2$O$_3$ Nanorods on Hollow Porous Carbon Nanofibers: A Free-Standing Electrode for Solid-State Symmetric Supercapacitors. *Chem. Eng. J.* **2022**, *429*, 132345. [CrossRef]
106. Yu, D.-G. Preface. *Curr. Drug Deliv.* **2021**, *18*, 2–3. [CrossRef]
107. Wang, M.; Li, D.; Li, J.; Li, S.; Chen, Z.; Yu, D.-G.; Liu, Z.; Guo, J.Z. Electrospun Janus Zein–PVP Nanofibers Provide a Two-Stage Controlled Release of Poorly Water-Soluble Drugs. *Mater. Des.* **2020**, *196*, 109075. [CrossRef]
108. Luraghi, A.; Peri, F.; Moroni, L. Electrospinning for Drug Delivery Applications: A Review. *J. Control. Release* **2021**, *334*, 463–484. [CrossRef]
109. Taskin, M.B.; Ahmad, T.; Wistlich, L.; Meinel, L.; Schmitz, M.; Rossi, A.; Groll, J. Bioactive Electrospun Fibers: Fabrication Strategies and a Critical Review of Surface-Sensitive Characterization and Quantification. *Chem. Rev.* **2021**, *121*, 11194–11237. [CrossRef] [PubMed]
110. Jurak, M.; Wiącek, A.E.; Ładniak, A.; Przykaza, K.; Szafran, K. What Affects the Biocompatibility of Polymers? *Adv. Colloid Interface Sci.* **2021**, *294*, 102451. [CrossRef]
111. Zhou, K.; Wang, M.; Zhou, Y.; Sun, M.; Xie, Y.; Yu, D.-G. Comparisons of Antibacterial Performances between Electrospun Polymer@drug Nanohybrids with Drug-Polymer Nanocomposites. *Adv. Compos. Hybrid Mater.* **2022**, *5*. [CrossRef]
112. Tuğcu-Demiröz, F.; Saar, S.; Kara, A.A.; Yıldız, A.; Tunçel, E.; Acartürk, F. Development and Characterization of Chitosan Nanoparticles Loaded Nanofiber Hybrid System for Vaginal Controlled Release of Benzydamine. *Eur. J. Pharm. Sci.* **2021**, *161*, 105801. [CrossRef]
113. He, H.; Wu, M.; Zhu, J.; Yang, Y.; Ge, R.; Yu, D.-G. Engineered Spindles of Little Molecules Around Electrospun Nanofibers for Biphasic Drug Release. *Adv. Fiber Mater.* **2021**, *3*, 1–13. [CrossRef]
114. Li, B.; Xia, X.; Chen, J.; Xia, D.; Xu, R.; Zou, X.; Wang, H.; Liang, C. Paclitaxel-Loaded Lignin Particle Encapsulated into Electrospun PVA/PVP Composite Nanofiber for Effective Cervical Cancer Cell Inhibition. *Nanotechnology* **2020**, *32*, 015101. [CrossRef] [PubMed]
115. Ashrafizadeh, M.; Mirzaei, S.; Gholami, M.H.; Hashemi, F.; Zabolian, A.; Raei, M.; Hushmandi, K.; Zarrabi, A.; Voelcker, N.H.; Aref, A.R.; et al. Hyaluronic Acid-Based Nanoplatforms for Doxorubicin: A Review of Stimuli-Responsive Carriers, Co-Delivery and Resistance Suppression. *Carbohydr. Polym.* **2021**, *272*, 118491. [CrossRef]
116. Luo, H.; Zhang, Y.; Yang, Z.; Zuo, G.; Zhang, Q.; Yao, F.; Wan, Y. Encapsulating Doxorubicin-Intercalated Lamellar Nanohydroxyapatite into PLGA Nanofibers for Sustained Drug Release. *Curr. Appl. Phys.* **2019**, *19*, 1204–1210. [CrossRef]
117. Izci, M.; Maksoudian, C.; Manshian, B.B.; Soenen, S.J. The Use of Alternative Strategies for Enhanced Nanoparticle Delivery to Solid Tumors. *Chem. Rev.* **2021**, *121*, 1746–1803. [CrossRef] [PubMed]
118. Xu, L.; Li, W.; Sadeghi-Soureh, S.; Amirsaadat, S.; Pourpirali, R.; Alijani, S. Dual Drug Release Mechanisms through Mesoporous Silica Nanoparticle/Electrospun Nanofiber for Enhanced Anticancer Efficiency of Curcumin. *J. Biomed. Mater. Res. Part A* **2021**, 1–15. [CrossRef]
119. Samadzadeh, S.; Babazadeh, M.; Zarghami, N.; Pilehvar-Soltanahmadi, Y.; Mousazadeh, H. An Implantable Smart Hyperthermia Nanofiber with Switchable, Controlled and Sustained Drug Release: Possible Application in Prevention of Cancer Local Recurrence. *Mater. Sci. Eng. C* **2021**, *118*, 111384. [CrossRef]
120. Croitoru, A.-M.; Karaçelebi, Y.; Saatcioglu, E.; Altan, E.; Ulag, S.; Aydoğan, H.K.; Sahin, A.; Motelica, L.; Oprea, O.; Tihauan, B.-M.; et al. Electrically Triggered Drug Delivery from Novel Electrospun Poly(Lactic Acid)/Graphene Oxide/Quercetin Fibrous Scaffolds for Wound Dressing Applications. *Pharmaceutics* **2021**, *13*, 957. [CrossRef]
121. Li, D.; Wang, M.; Song, W.-L.; Yu, D.-G.; Bligh, S.W.A. Electrospun Janus Beads-On-A-String Structures for Different Types of Controlled Release Profiles of Double Drugs. *Biomolecules* **2021**, *11*, 635. [CrossRef]
122. Gupta, N.; Kamath, S.M.; Rao, S.K.; Jaison, D.; Patil, S.; Gupta, N.; Arunachalam, K.D. Kaempferol Loaded Albumin Nanoparticles and Dexamethasone Encapsulation into Electrospun Polycaprolactone Fibrous Mat–Concurrent Release for Cartilage Regeneration. *J. Drug Deliv. Sci. Technol.* **2021**, *64*, 102666. [CrossRef]
123. Chen, K.; Pan, H.; Ji, D.; Li, Y.; Duan, H.; Pan, W. Curcumin-Loaded Sandwich-like Nanofibrous Membrane Prepared by Electrospinning Technology as Wound Dressing for Accelerate Wound Healing. *Mater. Sci. Eng. C* **2021**, *127*, 112245. [CrossRef] [PubMed]
124. Bardoňová, L.; Kotzianová, A.; Skuhrovcová, K.; Židek, O.; Bártová, T.; Kulhánek, J.; Hanová, T.; Mamulová Kutláková, K.; Vágnerová, H.; Krpatová, V.; et al. Antimicrobial Nanofibrous Mats with Controllable Drug Release Produced from Hydrophobized Hyaluronan. *Carbohydr. Polym.* **2021**, *267*, 118225. [CrossRef] [PubMed]
125. Wang, Z.; Song, X.; Cui, Y.; Cheng, K.; Tian, X.; Dong, M.; Liu, L. Silk Fibroin H-Fibroin/Poly(ε-Caprolactone) Core-Shell Nanofibers with Enhanced Mechanical Property and Long-Term Drug Release. *J. Colloid Interface Sci.* **2021**, *593*, 142–151. [CrossRef] [PubMed]
126. Darvishi, E.; Kahrizi, D.; Arkan, E.; Hosseinabadi, S.; Nematpour, N. Preparation of Bio-Nano Bandage from Quince Seed Mucilage/ZnO Nanoparticles and Its Application for the Treatment of Burn. *J. Mol. Liq.* **2021**, *339*, 116598. [CrossRef]
127. Liu, W.; Zhang, H.; Zhang, W.; Wang, M.; Li, J.; Zhang, Y.; Li, H. Surface Modification of a Polylactic Acid Nanofiber Membrane by Zeolitic Imidazolate Framework-8 from Secondary Growth for Drug Delivery. *J. Mater. Sci.* **2020**, *55*, 15275–15287. [CrossRef]
128. Liu, Y.; Chen, X.; Yu, D.-G.; Liu, H.; Liu, Y.; Liu, P. Electrospun PVP-Core/PHBV-Shell Fibers to Eliminate Tailing Off for an Improved Sustained Release of Curcumin. *Mol. Pharm.* **2021**, *18*, 4170–4178. [CrossRef]

129. Banerjee, A.; Jariwala, T.; Baek, Y.-K.; To, D.T.H.; Tai, Y.; Liu, J.; Park, H.; Myung, N.V.; Nam, J. Magneto- and Opto-Stimuli Responsive Nanofibers as a Controlled Drug Delivery System. *Nanotechnology* **2021**, *32*, 505101. [CrossRef]

130. Do Pazo-Oubiña, F.; Alorda-Ladaria, B.; Gomez-Lobon, A.; Boyeras-Vallespir, B.; Santandreu-Estelrich, M.M.; Martorell-Puigserver, C.; Gomez-Zamora, M.; Ventayol-Bosch, P.; Delgado-Sanchez, O. Thermolabile Drug Storage in an Ambulatory Setting. *Sci. Rep.* **2021**, *11*, 5959. [CrossRef]

131. Sedláček, O.; Černoch, P.; Kučka, J.; Konefal, R.; Štěpánek, P.; Vetrík, M.; Lodge, T.P.; Hrubý, M. Thermoresponsive Polymers for Nuclear Medicine: Which Polymer Is the Best? *Langmuir* **2016**, *32*, 6115–6122. [CrossRef]

132. Yang, J.; Wang, K.; Yu, D.-G.; Yang, Y.; Bligh, S.W.A.; Williams, G.R. Electrospun Janus Nanofibers Loaded with a Drug and Inorganic Nanoparticles as an Effective Antibacterial Wound Dressing. *Mater. Sci. Eng. C* **2020**, *111*, 110805. [CrossRef]

133. Bakhsheshi-Rad, H.R.; Ismail, A.F.; Aziz, M.; Akbari, M.; Hadisi, Z.; Khoshnava, S.M.; Pagan, E.; Chen, X. Co-Incorporation of Graphene Oxide/Silver Nanoparticle into Poly-L-Lactic Acid Fibrous: A Route toward the Development of Cytocompatible and Antibacterial Coating Layer on Magnesium Implants. *Mater. Sci. Eng. C* **2020**, *111*, 110812. [CrossRef]

134. Yaqoob, A.A.; Ahmad, H.; Parveen, T.; Ahmad, A.; Oves, M.; Ismail, I.M.I.; Qari, H.A.; Umar, K.; Mohamad Ibrahim, M.N. Recent Advances in Metal Decorated Nanomaterials and Their Various Biological Applications: A Review. *Front. Chem.* **2020**, *8*, 341. [CrossRef]

135. Ibrahim, H.M.; Reda, M.M.; Klingner, A. Preparation and Characterization of Green Carboxymethylchitosan (CMCS)–Polyvinyl Alcohol (PVA) Electrospun Nanofibers Containing Gold Nanoparticles (AuNPs) and Its Potential Use as Biomaterials. *Int. J. Biol. Macromol.* **2020**, *151*, 821–829. [CrossRef] [PubMed]

136. Nguyen, T.H.A.; Nguyen, V.-C.; Phan, T.N.H.; Le, V.T.; Vasseghian, Y.; Trubitsyn, M.A.; Nguyen, A.-T.; Chau, T.P.; Doan, V.-D. Novel Biogenic Silver and Gold Nanoparticles for Multifunctional Applications: Green Synthesis, Catalytic and Antibacterial Activity, and Colorimetric Detection of Fe(III) Ions. *Chemosphere* **2022**, *287*, 132271. [CrossRef]

137. Li, Z.; Tian, C.; Jiao, D.; Li, J.; Li, Y.; Zhou, X.; Zhao, H.; Zhao, Y.; Han, X. Synergistic Effects of Silver Nanoparticles and Cisplatin in Combating Inflammation and Hyperplasia of Airway Stents. *Bioact. Mater.* **2022**, *9*, 266–280. [CrossRef] [PubMed]

138. Hu, X.; He, J.; Zhu, L.; Machmudah, S.; Wahyudiono; Kanda, H.; Goto, M. Synthesis of Hollow PVP/Ag Nanoparticle Composite Fibers via Electrospinning under a Dense CO_2 Environment. *Polymers* **2022**, *14*, 89. [CrossRef] [PubMed]

139. Zhu, Z.; Jin, L.; Yu, F.; Wang, F.; Weng, Z.; Liu, J.; Han, Z.; Wang, X. ZnO/CPAN Modified Contact Lens with Antibacterial and Harmful Light Reduction Capabilities. *Adv. Healthc. Mater.* **2021**, *10*, 2100259. [CrossRef]

140. Zhang, R.; Jiang, G.; Gao, Q.; Wang, X.; Wang, Y.; Xu, X.; Yan, W.; Shen, H. Sprayed Copper Peroxide Nanodots for Accelerating Wound Healing in a Multidrug-Resistant Bacteria Infected Diabetic Ulcer. *Nanoscale* **2021**, *13*, 15937–15951. [CrossRef]

141. Shi, J.; Li, J.; Wang, Y.; Zhang, C.Y. TiO₂-Based Nanosystem for Cancer Therapy and Antimicrobial Treatment: A Review. *Chem. Eng. J.* **2021**, *431*, 133714. [CrossRef]

142. Villarreal-Gómez, L.J.; Pérez-González, G.L.; Bogdanchikova, N.; Pestryakov, A.; Nimaev, V.; Soloveva, A.; Cornejo-Bravo, J.M.; Toledaño-Magaña, Y. Antimicrobial Effect of Electrospun Nanofibers Loaded with Silver Nanoparticles: Influence of Ag Incorporation Method. *J. Nanomater.* **2021**, *2021*, e9920755. [CrossRef]

143. Kakakhel, M.A.; Wu, F.; Sajjad, W.; Zhang, Q.; Khan, I.; Ullah, K.; Wang, W. Long-Term Exposure to High-Concentration Silver Nanoparticles Induced Toxicity, Fatality, Bioaccumulation, and Histological Alteration in Fish (Cyprinus Carpio). *Environ. Sci. Eur.* **2021**, *33*, 14. [CrossRef]

144. Lemraski, E.G.; Jahangirian, H.; Dashti, M.; Khajehali, E.; Sharafinia, S.; Rafiee-Moghaddam, R.; Webster, T.J. Antimicrobial Double-Layer Wound Dressing Based on Chitosan/Polyvinyl Alcohol/Copper: In Vitro and in Vivo Assessment. *Int. J. Nanomed.* **2021**, *16*, 223–235. [CrossRef] [PubMed]

145. Wang, Q.; Feng, Y.; He, M.; Zhao, W.; Qiu, L.; Zhao, C. A Hierarchical Janus Nanofibrous Membrane Combining Direct Osteogenesis and Osteoimmunomodulatory Functions for Advanced Bone Regeneration. *Adv. Funct. Mater.* **2021**, *31*, 2008906. [CrossRef]

146. Saatchi, A.; Arani, A.R.; Moghanian, A.; Mozafari, M. Cerium-Doped Bioactive Glass-Loaded Chitosan/Polyethylene Oxide Nanofiber with Elevated Antibacterial Properties as a Potential Wound Dressing. *Ceram. Int.* **2021**, *47*, 9447–9461. [CrossRef]

147. Norouzi, M.A.; Montazer, M.; Harifi, T.; Karimi, P. Flower Buds like PVA/ZnO Composite Nanofibers Assembly: Antibacterial, in Vivo Wound Healing, Cytotoxicity and Histological Studies. *Polym. Test.* **2021**, *93*, 106914. [CrossRef]

148. Khan, A.U.R.; Huang, K.; Jinzhong, Z.; Zhu, T.; Morsi, Y.; Aldalbahi, A.; El-Newehy, M.; Yan, X.; Mo, X. Exploration of the Antibacterial and Wound Healing Potential of a PLGA/Silk Fibroin Based Electrospun Membrane Loaded with Zinc Oxide Nanoparticles. *J. Mater. Chem. B* **2021**, *9*, 1452–1465. [CrossRef]

149. Huang, C.; Dong, J.; Zhang, Y.; Chai, S.; Wang, X.; Kang, S.; Yu, D.; Wang, P.; Jiang, Q. Gold Nanoparticles-Loaded Polyvinylpyrrolidone/Ethylcellulose Coaxial Electrospun Nanofibers with Enhanced Osteogenic Capability for Bone Tissue Regeneration. *Mater. Des.* **2021**, *212*, 110240. [CrossRef]

150. Chen, Z.; Xiao, H.; Zhang, H.; Xin, Q.; Zhang, H.; Liu, H.; Wu, M.; Zuo, L.; Luo, J.; Guo, Q.; et al. Heterogenous Hydrogel Mimicking the Osteochondral ECM Applied to Tissue Regeneration. *J. Mater. Chem. B* **2021**, *9*, 8646–8658. [CrossRef] [PubMed]

151. Rahmati, M.; Mills, D.K.; Urbanska, A.M.; Saeb, M.R.; Venugopal, J.R.; Ramakrishna, S.; Mozafari, M. Electrospinning for Tissue Engineering Applications. *Prog. Mater. Sci.* **2021**, *117*, 100721. [CrossRef]

152. Lu, Z.; Wang, W.; Zhang, J.; Bártolo, P.; Gong, H.; Li, J. Electrospun Highly Porous Poly(L-Lactic Acid)-Dopamine-SiO₂ Fibrous Membrane for Bone Regeneration. *Mater. Sci. Eng. C* **2020**, *117*, 111359. [CrossRef]

153. Stahl, A.; Yang, Y.P. Regenerative Approaches for the Treatment of Large Bone Defects. *Tissue Eng. Part B* **2021**, *27*, 539–547. [CrossRef]
154. Hassani, A.; Khoshfetrat, A.B.; Rahbarghazi, R.; Sakai, S. Collagen and Nano-Hydroxyapatite Interactions in Alginate-Based Microcapsule Provide an Appropriate Osteogenic Microenvironment for Modular Bone Tissue Formation. *Carbohydr. Polym.* **2022**, *277*, 118807. [CrossRef] [PubMed]
155. Kermani, F.; Mollazadeh, S.; Kargozar, S.; Vahdati Khakhi, J. Improved Osteogenesis and Angiogenesis of Theranostic Ions Doped Calcium Phosphates (CaPs) by a Simple Surface Treatment Process: A State-of-the-Art Study. *Mater. Sci. Eng. C* **2021**, *124*, 112082. [CrossRef] [PubMed]
156. Ma, K.; Liao, C.; Huang, L.; Liang, R.; Zhao, J.; Zheng, L.; Su, W. Electrospun PCL/MoS₂ Nanofiber Membranes Combined with NIR-Triggered Photothermal Therapy to Accelerate Bone Regeneration. *Small* **2021**, *17*, 2104747. [CrossRef] [PubMed]
157. Lin, Z.; Shen, D.; Zhou, W.; Zheng, Y.; Kong, T.; Liu, X.; Wu, S.; Chu, P.K.; Zhao, Y.; Wu, J.; et al. Regulation of Extracellular Bioactive Cations in Bone Tissue Microenvironment Induces Favorable Osteoimmune Conditions to Accelerate in Situ Bone Regeneration. *Bioact. Mater.* **2021**, *6*, 2315–2330. [CrossRef]
158. Zhang, Y.; Wang, P.; Mao, H.; Zhang, Y.; Zheng, L.; Yu, P.; Guo, Z.; Li, L.; Jiang, Q. PEGylated Gold Nanoparticles Promote Osteogenic Differentiation in in Vitro and in Vivo Systems. *Mater. Des.* **2021**, *197*, 109231. [CrossRef]
159. Song, H.; Zhang, Y.; Zhang, Z.; Xiong, S.; Ma, X.; Li, Y. Hydroxyapatite/NELL-1 Nanoparticles Electrospun Fibers for Osteoinduction in Bone Tissue Engineering Application. *Int. J. Nanomed.* **2021**, *16*, 4321–4332. [CrossRef]
160. Liu, W.; Bi, W.; Sun, Y.; Wang, L.; Yu, X.; Cheng, R.; Yu, Y.; Cui, W. Biomimetic Organic-Inorganic Hybrid Hydrogel Electrospinning Periosteum for Accelerating Bone Regeneration. *Mater. Sci. Eng. C* **2020**, *110*, 110670. [CrossRef] [PubMed]
161. Shi, Y.-H.; Li, X.-Y.; Zhang, W.-D.; Li, H.-H.; Wang, S.-G.; Wu, X.-L.; Su, Z.-M.; Zhang, J.-P.; Sun, H.-Z. Micron-Scaled MoS₂/N-C Particles with Embedded Nano-MoS₂: A High-Rate Anode Material for Enhanced Lithium Storage. *Appl. Surf. Sci.* **2019**, *486*, 519–526. [CrossRef]
162. Wu, S.; Wang, J.; Jin, L.; Li, Y.; Wang, Z. Effects of Polyacrylonitrile/MoS2 Composite Nanofibers on the Growth Behavior of Bone Marrow Mesenchymal Stem Cells. *ACS Appl. Nano Mater.* **2018**, *1*, 337–343. [CrossRef]
163. Kasper, M.; Ellenbogen, B.; Hardy, R.; Cydis, M.; Mojica-Santiago, J.; Afridi, A.; Spearman, B.S.; Singh, I.; Kuliasha, C.A.; Atkinson, E.; et al. Development of a Magnetically Aligned Regenerative Tissue-Engineered Electronic Nerve Interface for Peripheral Nerve Applications. *Biomaterials* **2021**, *279*, 121212. [CrossRef] [PubMed]
164. Chen, X.; Ge, X.; Qian, Y.; Tang, H.; Song, J.; Qu, X.; Yue, B.; Yuan, W.-E. Electrospinning Multilayered Scaffolds Loaded with Melatonin and Fe₃O₄ Magnetic Nanoparticles for Peripheral Nerve Regeneration. *Adv. Funct. Mater.* **2020**, *30*, 2004537. [CrossRef]
165. Morsink, M.; Severino, P.; Luna-Ceron, E.; Hussain, M.A.; Sobahi, N.; Shin, S.R. Effects of Electrically Conductive Nano-Biomaterials on Regulating Cardiomyocyte Behavior for Cardiac Repair and Regeneration. *Acta Biomater.* **2021**, *126*, S17427061(21)007716. [CrossRef]
166. Zhao, G.; Zhang, X.; Li, B.; Huang, G.; Xu, F.; Zhang, X. Solvent-Free Fabrication of Carbon Nanotube/Silk Fibroin Electrospun Matrices for Enhancing Cardiomyocyte Functionalities. *ACS Biomater. Sci. Eng.* **2020**, *6*, 1630–1640. [CrossRef]
167. He, C.; Liu, X.; Zhou, Z.; Liu, N.; Ning, X.; Miao, Y.; Long, Y.; Wu, T.; Leng, X. Harnessing Biocompatible Nanofibers and Silver Nanoparticles for Wound Healing: Sandwich Wound Dressing versus Commercial Silver Sulfadiazine Dressing. *Mater. Sci. Eng. C* **2021**, *128*, 112342. [CrossRef]
168. Talikowska, M.; Fu, X.; Lisak, G. Application of Conducting Polymers to Wound Care and Skin Tissue Engineering: A Review. *Biosens. Bioelectron.* **2019**, *135*, 50–63. [CrossRef] [PubMed]
169. Ji, X.; Li, R.; Liu, G.; Jia, W.; Sun, M.; Liu, Y.; Luo, Y.; Cheng, Z. Phase Separation-Based Electrospun Janus Nanofibers Loaded with Rana Chensinensis Skin Peptides/Silver Nanoparticles for Wound Healing. *Mater. Des.* **2021**, *207*, 109864. [CrossRef]
170. Gu, Y.; Son, S.U.; Li, T.; Tan, B. Low-Cost Hypercrosslinked Polymers by Direct Knitting Strategy for Catalytic Applications. *Adv. Funct. Mater.* **2021**, *31*, 2008265. [CrossRef]
171. Tan, L.; Tan, B. Hypercrosslinked Porous Polymer Materials: Design, Synthesis, and Applications. *Chem. Soc. Rev.* **2017**, *46*, 3322–3356. [CrossRef] [PubMed]
172. Song, W.; Zhang, Y.; Varyambath, A.; Kim, I. Guided Assembly of Well-Defined Hierarchical Nanoporous Polymers by Lewis Acid–Base Interactions. *ACS Nano* **2019**, *13*, 11753–11769. [CrossRef]
173. Song, W.; Zhang, Y.; Yu, D.-G.; Tran, C.H.; Wang, M.; Varyambath, A.; Kim, J.; Kim, I. Efficient Synthesis of Folate-Conjugated Hollow Polymeric Capsules for Accurate Drug Delivery to Cancer Cells. *Biomacromolecules* **2021**, *22*, 732–742. [CrossRef]
174. Petre, D.G.; Leeuwenburgh, S.C.G. The Use of Fibers in Bone Tissue Engineering. *Tissue Eng. Part B Rev.* **2021**, *27*. [CrossRef]
175. Wei, C.; Feng, Y.; Che, D.; Zhang, J.; Zhou, X.; Shi, Y.; Wang, L. Biomaterials in Skin Tissue Engineering. *Int. J. Polym. Mater. Polym. Biomater.* **2021**, 1–19. [CrossRef]
176. Yahya, E.B.; Amirul, A.A.; Abdul Khalil, H.P.S.; Olaiya, N.G.; Iqbal, M.O.; Jummaat, F.; Atty Sofea, A.K.; Adnan, A.S. Insights into the Role of Biopolymer Aerogel Scaffolds in Tissue Engineering and Regenerative Medicine. *Polymers* **2021**, *13*, 1612. [CrossRef] [PubMed]
177. Hsu, Y.-J.; Wei, S.-Y.; Lin, T.-Y.; Fang, L.; Hsieh, Y.-T.; Chen, Y.-C. A Strategy to Engineer Vascularized Tissue Constructs by Optimizing and Maintaining the Geometry. *Acta Biomater.* **2022**, *138*, 254–272. [CrossRef]
178. Yu, D.G.; Wang, M.; Ge, R. Strategies for Sustained Drug Release from Electrospun Multi-layer Nanostructures. *WIRES Nanomed. Nanobi.* **2021**, *14*, e1772. [CrossRef] [PubMed]
179. Kundrat, V.; Vykoukal, V.; Moravec, Z.; Simonikova, L.; Novotny, K.; Pinkas, J. Preparation of Polycrystalline Tungsten Nanofibers by Needleless Electrospinning. *J. Alloys Compd.* **2022**, *900*, 163542. [CrossRef]

 *polymers*

Article

Investigation of Mechanical, Chemical, and Antibacterial Properties of Electrospun Cellulose-Based Scaffolds Containing Orange Essential Oil and Silver Nanoparticles

Duy-Nam Phan [1,*], Muhammad Qamar Khan [2,*], Van-Chuc Nguyen [3], Hai Vu-Manh [1], Anh-Tuan Dao [1], Phan Thanh Thao [1], Ngoc-Mai Nguyen [3], Van-Tuan Le [4], Azeem Ullah [5], Muzamil Khatri [6] and Ick-Soo Kim [5,*]

1 School of Textile-Leather and Fashion, Hanoi University of Science and Technology, 1 Dai Co Viet, Hanoi 10000, Vietnam; hai.vumanh@hust.edu.vn (H.V.-M.); tuan.daoanh@hust.edu.vn (A.-T.D.); thao.phanthanh@hust.edu.vn (P.T.T.)
2 Department of Textile and Clothing, Faculty of Textile Engineering and Technology, National Textile University, Karachi Campus, Karachi 74900, Pakistan
3 School of Chemical Engineering, Hanoi University of Science and Technology, 1 Dai Co Viet, Hanoi 10000, Vietnam; chucnguyenvan85@gmail.com (V.-C.N.); mai.nguyenngoc@hust.edu.vn (N.-M.N.)
4 School of Mechanical Engineering, Hanoi University of Science and Technology, 1 Dai Co Viet, Hanoi 10000, Vietnam; tuan.levan@hust.edu.vn
5 Nano Fusion Technology Research Group, Institute for Fiber Engineering (IFES), Interdisciplinary Cluster for Cutting Edge Research (ICCER), Shinshu University, Tokida 3-15-1, Ueda 386-8567, Nagano, Japan; 08tex101@gmail.com
6 Department of Chemistry and Materials, Faculty of Textile Science and Technology, Shinshu University, Tokida 3-15-1, Ueda 386-8567, Nagano, Japan; muzamilkhatri@gmail.com
* Correspondence: nam.phanduy@hust.edu.vn (D.-N.P.); qamarkhan154@gmail.com (M.Q.K.); kim@shinshu-u.ac.jp (I.-S.K.)

Citation: Phan, D.-N.; Khan, M.Q.; Nguyen, V.-C.; Vu-Manh, H.; Dao, A.-T.; Thanh Thao, P.; Nguyen, N.-M.; Le, V.-T.; Ullah, A.; Khatri, M.; et al. Investigation of Mechanical, Chemical, and Antibacterial Properties of Electrospun Cellulose-Based Scaffolds Containing Orange Essential Oil and Silver Nanoparticles. *Polymers* 2022, 14, 85. https://doi.org/10.3390/polym14010085

Academic Editor: Tao-Hsing Chen

Received: 16 November 2021
Accepted: 23 December 2021
Published: 27 December 2021

Publisher's Note: MDPI stays neutral with regard to jurisdictional claims in published maps and institutional affiliations.

Abstract: This study demonstrated a controllable release properties and synergistic antibacterial actions between orange essential oil (OEO) and silver nanoparticles (AgNPs) incorporated onto cellulose (CL) nanofibers. The preparation of AgNPs attached on CL nanofibers was conducted through multiple processes including the deacetylation process to transform cellulose acetate (CA) nanofibers to CL nanofibers, the in situ synthesis of AgNPs, and the coating of as-prepared silver composite CL nanofibers using OEO solutions with two different concentrations. The success of immobilization of AgNPs onto the surface of CL nanofibers and the incorporation of OEO into the polymer matrix was confirmed by SEM-EDS, TEM, XRD, and FT-IR characterizations. The tensile strength, elongation at break, and Young's modulus of the nanofibers after each step of treatment were recorded and compared to pristine CA nanofibers. The high antibacterial activities of AgNPs and OEO were assessed against Gram-positive *B. subtilis* and Gram-negative *E. coli* microorganisms. The combined effects of two antimicrobials, AgNPs and OEO, were distinctively recognized against *E. coli*.

Keywords: cellulose nanofiber; silver nanoparticle; electrospinning; orange essential oil; antibacterial activity

1. Introduction

Cellulose is regarded as the most plentiful renewable bio-polymer on earth with the chemical structure of d-glucopyranosyl units joined by β-1,4-glycosidic bonds. Cellulose can be harvested through plants, wood, marine organisms, and biosynthesis for the manufacture of cellulose-based materials such as fibers, films, hydrogels, aerogels, and composites. Cellulose presents crystalline regions where the chains are tightly packed and with uniform cellulose molecular arrangements and amorphous domains with loose hydrogen bonding [1,2]. The rich hydroxyl groups on cellulose provide chemical modification to acquire cellulose derivatives and play a crucial role on controlling physicochemical properties of cellulose [3,4]. Cellulose nanofibers with excellent mechanical properties, highly thermostable due to the hydrogen bond intra macromolecules, have been used for biomedical applications, filtration, and food packaging [5,6].

Nanofibers are one-dimensional materials with promising applications such as in electronics, energy storage cells, catalysis, and filtering membrane [7,8]. The combination of polymers, metal, metal oxides, and other substances grants nanofibers multifunctional properties. Among many methods for nanofiber manufacture, electrospinning is a straight-forward and versatile technique to fabricate nanofibers relying on electrostatic repulsion between charged polymeric solution and collector [9–11]. The method has been used to produce nanofibers or composite nanofibers with the diameters of tens to a few hundred nanometers. CL nanofibers fabricated using electrospinning method showcase Young's modulus in the range of 120 to 170 GPa, 150 to 170 m^2/g specific surface areas, high crystallinities, and thermal stability at temperatures up to 300 °C [12].

AgNPs have been extensively studied and used in the field of medicine in recent decades due to their broad spectrum of antibacterial, physical, chemical, and biological properties [13]. AgNPs have diameters in the range of few to 100 nm with a large surface-to-volume ratio, with low cytotoxicity for humans [14]. The mechanisms of silver antibacterial effects can be summarized as ion release, increased cellular oxidative stress, reactions with sulfur-containing cell membranes and phosphorus in DNA, and attraction of Ag to the negatively charged cell membrane of microorganisms [15,16]. Moreover, AgNPs are biocompatible with keratinocyte and fibroblasts and also display anti-inflammatory properties, even at very low concentrations [17].

Sweet orange is one of the most popular citrus fruits with fragrant taste and health promotion chemical compounds, largely available in many countries [18]. However, the peel is mostly discarded as biowaste or landfilling, which can result in environmental issues and is not economically desirable [19]. Orange essential oil has bioactive components that exhibit antioxidative, antimicrobial, and antiseptic activities against different pathogens and bacterial strains [20]. The use of essential oils (EO) in food and cosmetic industries is promoted due to health concerns of chemical substances [21].

Essential oil and silver-containing cellulosic scaffolds have valuable applications in drug delivery, release systems, medicine, and food packaging. Moreover, there are growing environmental concerns for synthetic polymers; therefore, biopolymers as cellulose and EO used as biodegradable scaffolds show great interest for the food industry and facial masks due to crack resistance, high porosity, and controllable compound release. In order to evaluate the feasibility of fabricating cellulose nanofibrous webs for applications related to antibacterial properties such as masks, wound dressings, food packaging, and protective clothing, we engineered AgNPs and OEO to the CL nanofiber scaffolds. The OEO contents were changed to explore the influence of OEO as coating material in CL composite nanofibers regarding chemical, mechanical, and antibacterial activities.

2. Materials and Methods

2.1. Materials

Cellulose acetate (average Mn = ≈50,000) was purchased from Sigma-Aldrich Chemical Co., Ltd. (St. Louis, MO, USA). Sodium hydroxide (NaOH, 97%), diethyl ether (99%), and silver nitrate ($AgNO_3$, 99.8%) were acquired from Shanghai Aladdin Bio-Chem Technology Co., Ltd. (Shanghai, China). Acetone (99.5%) and N,N-dimethylformamide (DMF, 99.8%) were purchased from Wako Pure Chemical Industries, Ltd. (Osaka, Japan). Sodium borohydride ($NaBH_4$, 95%) was obtained from BDH chemicals Ltd. (Poole, UK). Orange essential oil was extracted from the peel of Vietnam's sweet orange (citrus sinensis) and purchased from local company. All chemicals were used as received without further purification. *Bacillus subtilis* ATTC 6633 and *Escherichia coli* ATCC 25922 were obtained from Microbiologics, Inc (Saint Cloud, MN, USA).

2.2. Silver and Orange Essential Oil Integrated in Cellulose Nanofibers

Cellulose nanofibers were prepared through two steps, electrospinning CA nanofibers and deacetylation process to convert CA nanofibers to CL nanofibers (Scheme 1). To prepare CA solution, we dissolved CA polymer in a bi-solvent system of acetone/DMF with the volume ratio of 1:1 and polymer weight percentage of 15 wt %. The solution was vigorously stirred for 12 h to dissolve the solute completely before being spun. The resultant CA solution was loaded into a 5 mL syringe with a metallic capillary needle of 0.6 mm internal diameter. The distance between the tip of the needle and the collector was fixed at 15 cm and the supplied voltage was 20 kV. The power supplier (HVU-30P100) has voltage and current capacity of 30 kV and 100 μA, respectively, and was purchased from MECC Co., Ltd., Fukuoka, Japan. The CA nanofibrous mats were collected after 24 h electrospinning. Afterward, the CA nanofibers were deacetylated by NaOH 0.01 M for 48 h. The acquired CL nanofibers were washed thoroughly with deionized water before being dried at 40 °C for 8 h, and eventually the immobilization process of AgNPs onto the CL nanofibrous webs occurred.

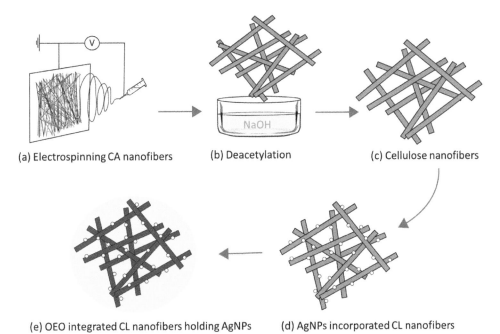

(a) Electrospinning CA nanofibers (b) Deacetylation (c) Cellulose nanofibers

(e) OEO integrated CL nanofibers holding AgNPs (d) AgNPs incorporated CL nanofibers

Scheme 1. Schematic representation of the fabrication of orange essential oil-integrated cellulose nanofibers containing silver nanoparticles.

To prepare AgNP-incorporated cellulose nanofibers, CL nanofibers were firstly immersed in 0.05 M AgNO₃ solution for 1 h. The AgNO₃-CL nanofibers were subsequently let to dry in an atmospheric environment for 24 h. The dried AgNO₃-CL nanofibers were then dipped in NaBH₄ 0.01 M solution for 1 h to let all chemical reactions occur completely; AgNO₃ was reduced to AgNPs and entrapped into the CL nanofibrous webs. The composite Ag-CL nanofibers were collected, washed with DI water several times to remove all residual chemicals, and dried in air for 24 h.

The OEO solutions were prepared with two concentrations of 5% and 10% (v/v) in ethanol. CL nanofibers were soaked in 10% OEO solution; the sample was denoted as OEO-CL nanofibers. Ag-CL nanofibers were soaked in 5% OEO and 10% OEO solutions for 1 h, and the collected samples were indicated as Ag-OEO5%-CL and Ag-OEO10%-CL nanofibers.

2.3. Characterizations

The qualitative chemical composition of OEO was identified using a gas chromatograph coupled with a single quadrupole mass spectrometry (Thermo Scientific model TRACE 1310 Gas Chromatograph with an ISQ 7000 Single Quadrupole Mass Spectrometer, GC-SQ-MS). The extract was prepared in diethyl ether. The experiment was performed by using a Perkin Elmer Clarus 500 gas chromatography equipped with an Elite-5ms capillary column (30 m × 0.25 mm × 0.25 μm) as well as mass detector turbo mass gold of the company, which was set in EI mode. The carrier's gas flow was controlled at a rate of 1 mL/min. Simultaneously, the injector was managed at 250 °C. The oven temperature was programmed as follows: 50 °C at 5 °C/min (held for 5 min) to 140 °C at 7 °C/min and to 275 °C (held for 10 min).

The morphology and diameters of nanofibers were studied using scanning electron microscopy (SEM-JSM-6010LA, JOEL, Tokyo, Japan) at an accelerating voltage of 15 kV. The surface of the nanofibers was coated with platinum (Pt) in a sputtering chamber for 120 s at 30 mA. The fiber diameters with standard deviations were calculated using ImageJ application for 50 strands of fiber each sample. In addition, elemental analysis was studied on energy-dispersive X-ray spectroscopy (EDS).

The amorphous, crystalline, and semi-crystalline phases of all nanofibrous samples were investigated by X-ray diffraction instrument (XRD, Miniflex 300, Rigaku Co., Ltd., Tokyo, Japan) with Cu-Kα irradiation, operating at 30 kV and 500 mA, at the 2θ ranging from 5 ° to 80 °.

The characteristic vibration of functional groups present on the nanofibrous mats was analyzed using a Fourier transform infrared (FT-IR) spectrometer (Nicolet iS 50, FT-IR, Thermo, Waltham, MA, USA) in the range of 4000 to 400 cm^{-1} with accumulation over 20 scans.

The universal testing instrument (UTM, RTC-1250A, A&D Co., Ltd., Tokyo, Japan) was utilized to acquire stress–strain diagrams. The tensile test was performed five times for each sample, and the data were collected and used to calculate elongation at break, Young's modulus, and tensile strength. Specimens were cut into a rectangular shape of 70 × 10 mm (length × width). The thickness of each specimen was measured before the pull testing, and the cross-sectional area was calculated by multiplying the thickness and the width. The working conditions of the instrument were 10 mm/min extensional rate and 50 mm gauge length.

Nanofibrous specimens were prepared in pieces with the size of 2 × 2 cm for the water absorption test. In detail, all specimens were weighed and then immersed in deionized water for 1 h. Afterward, the specimens were removed and gently placed on filter papers to discharge all dripping water. The specimens were weighed again, and water absorption amount was calculated by gained mass over the original mass of the specimens.

$$\text{Water absorption (\%)} = \frac{(W_1 - W_0)}{W_0} \times 100$$

where W_1 is the specimen weight after water immersion and W_0 is the original weight of the specimen.

The silver release behaviors of silver-containing samples were examined in an aqueous environment using inductively coupled plasma mass spectroscopy (ICP-MS; Shimadzu ICPS-1000 IV; Shimadzu, Kyoto, Japan). In essence, 0.15 g of each silver-containing sample, Ag-CL, Ag-OEO5%-CL, and Ag-OEO10%-CL, were immersed in screw cap vials, each containing 50 mL deionized water, under gentle shaking and room temperature for 48 h. At various time intervals, 2 mL of liquid specimens were extracted and subjected to ICP-MS spectrometry to quantify silver concentrations. The release profiles were constructed by recording silver concentrations over a period of 48 h.

2.4. Antibacterial Assays

The antibacterial activities of CL as control sample, OEO-CL, Ag-CL, Ag-OEO5%-CL, and Ag-OEO10%-CL nanofibers were assessed against two bacterial strains, *E. coli* ATCC 25922 and *B. subtilis* ATTC 6633 in triplicate. The method was adopted from our previous publication [22]. In detail, *E. coli* was cultured in the LB liquid medium for 24 h at 37 °C, and *B. subtilis* was cultured likewise but at 30 °C. The cultured mediums were diluted by DI water to the concentration of 10^6 colony forming unit per milliliter—CFU/mL. For the disk diffusion test, 200 μL of bacteria suspension was smeared over an agar plate. The nanofibrous specimens were cut into 5 mm diameter round shape and placed on agar plates seeded with bacterial cells in log phase. All plates were incubated at optimal conditions for bacterial growth (37 °C for *E. coli* and 30 °C for *B. subtilis*) for 24 h. All disk diffusion test processes were conducted in triplicate to measure the diameters of halo zones and calculate the standard deviations.

3. Results and Discussion

3.1. Chemical Composition of the OEO

OEO was analyzed using GC-SQ-MS; the results showed the main constituents of OEO are d-limonene (64.33%), β-myrcene (8.84%), α-pinene (5.01%), α-phellandrene (2.82%), linalool (2.77%), decanal (1.89%), and octanal (1.28%) in 18.83, 17.29, 15.27, 16.72, 20.94, 24.2, and 17.68 (min), respectively (Table 1 and Figure S1). These seven compounds account for 86.94% of the total compounds identified. Limonene was also observed as the main component in the peel of sweet orange in several previous reports [23–25]. Among 36 volatile organic compounds detected, including monoterpenes, sesquiterpenes, aldehydes, alcohols, and esters, D-limonene was found to be the major contributor to the orange aroma.

Table 1. GC-SQ-MS-analyzed chemical compounds of OEO.

No.	Retention Time (min)	Compounds	Composition (%)
1	15.27	α-Pinene	**5.01**
2	16.72	α-Phellandrene	**2.82**
3	17.29	β-Myrcene	**8.84**
4	17.68	Octanal	**1.28**
5	18.83	D-Limonene	**64.33**
6	20.69	2-Carene	0.57
7	20.94	Linalool	**2.77**
8	21.45	p-Cymene	0.13
9	21.7	2-Caren-4-ol	0.34
10	22.14	cis-p-Mentha-2,8-dien-1-ol	0.93
11	22.27	(+)-(E)-Limonene oxide	0.42
12	22.61	Citronellal	0.37
13	23.26	1-Indanone, 4,5,6,7-tetrahydro	0.28
14	23.92	α-Terpineol	0.71
15	24.2	Decanal	**1.89**
16	24.73	2-Cyclohexen-1-ol, 2-methyl-5	0.63
17	25.08	Carveol	0.32
18	25.32	Neral	0.32
19	25.52	(−)-Carvone	0.74
20	26.14	Citral	0.76
21	26.45	2-Caren-10-al	0.31
22	28.15	Limonene oxide, trans-	0.16
23	29.37	Copaene	0.41
24	29.71	γ-Muurolene	0.46
25	29.88	Dodecanal	0.39

Table 1. *Cont.*

No.	Retention Time (min)	Compounds	Composition (%)
26	30.63	Caryophyllene	0.31
27	30.83	Copaene	0.43
28	32.47	Caryophyllene	0.82
29	32.71	Butylated hydroxytoluene	0.86
30	33.1	Cadina-1(10),4-diene	0.42
31	33.72	α-Acorenol	0.18
32	37.04	α-Longipinene	0.39
33	38.24	α-Sinensal	0.22
34	39.76	Nootkatone	0.15
35	42.35	m-Camphorene	0.16
36	43.05	1-Heptatriacotanol	0.1
Total			99.23

3.2. Morphology Study of Nanofiber Membranes

In Figure 1a, CA nanofibers are observed to be diverse in size; the average diameter value with deviation was found to be 409 ± 157 nm. With the concentration of 15 wt % CA in bi-solvent system of acetone and DMF, the polymer chains entangled enough with each other to form a smooth and bead-free nanofibers. After treatment with NaOH, CL nanofibers presented similar morphology with slightly smaller diameter, 399 ± 134 nm, due to the mercerization effects, which induced the nanofibers to be contracted in diameter according to previous studies [26,27]. Similarly, OEO-CL nanofibers did not present any significant difference compared to CL nanofibers, and the presence of OEO could not detected due to the evaporation of essential oil in the vacuum chamber of SEM equipment. In Figure 1d–f, the AgNPs could also be spotted as being attached to CL nanofibers. The average diameters of these nanofibers increased to 430 ± 109, 426 ± 128, and 432 ± 85 nm due to the treatment process of AgNP immobilization and the incorporation of OEO; AgNPs physically attached to the nanofibrous network structure. The morphology of Ag-OEO5%-CL and Ag-OEO10%-CL presented similar morphology compared to the Ag-CL sample because OEO was completely removed in the vacuum environment of the microscope chamber.

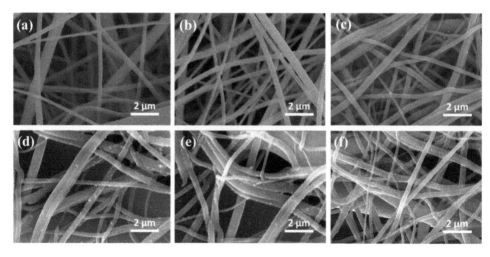

Figure 1. SEM images of (**a**) CA nanofibers, (**b**) CL nanofibers, (**c**) OEO-CL nanofibers, (**d**) Ag-CL nanofibers, (**e**) Ag-OEO5%-CL nanofibers, and (**f**) Ag-OEO10%-CL nanofibers.

3.3. FT-IR Spectral Analysis

Figure 2 presents the FT-IR spectra of pristine CA and CL nanofibers, and composite CL nanofibers after various steps of treatment. For CA spectrum, the strong absorbance peaks at 1735 cm^{-1}, 1365 cm^{-1}, and 1225 cm^{-1} corresponded to C = O vibration, C-CH$_3$ vibration, and C-O-C stretching vibration, respectively [28–30]. After the deacetylation reaction, the complete disappearance of the peaks at 1735 cm^{-1} and 1225 cm^{-1} and the raising band at 3400 cm^{-1} proved the successful conversion of CA to CL. The addition of OEO gave rise to the peaks at 2920 cm^{-1} and 2860 cm^{-1} assigned to -CH stretch, and 1735 cm^{-1} was attributed to carbonyl stretch of alkanes (Figure 2a) [31]. The peak at 872 cm^{-1} of AgNO$_3$-CL spectrum was assigned to -NO$_3$ bending, which can be referred to the penetration of silver nitrate into the cellulosic polymer matrix. This peak was removed after the redox reaction between AgNO$_3$ and NaBH$_4$ (Figure 2b), attesting to the conversion from silver salt to AgNPs.

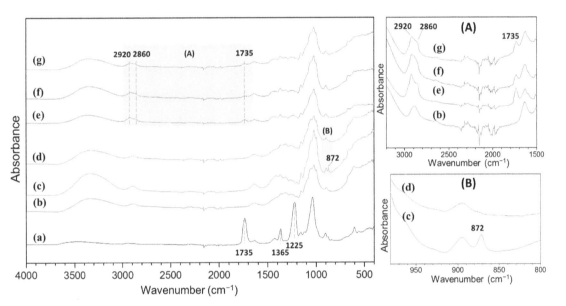

Figure 2. The FT-IR spectra of (**a**) CA, (**b**) CL, (**c**) AgNO$_3$-CL, (**d**) Ag-CL, (**e**) OEO-CL, (**f**) Ag-OEO5%-CL, and (**g**) Ag-OEO10%-CL; (**A**) magnified view of FTIR spectrum of OEO; and (**B**) magnified view of FTIR spectrum of AgNO$_3$.

3.4. X-ray Diffraction Study

The XRD patterns of CA and CL nanofibers, and AgNPs were identified using XRD technique. CA nanofibers exhibited two typical diffraction bands centered at 2θ of 9° and 22.5°, which corresponded to the semi-crystallinity of cellulose acetate polymer (Figure 3A(a)). The deacetylation process transformed CA to CL nanofibers completely, which showed typical bands of cellulose II. With the addition of OEO into the CL nanofibers, no characteristic bands or peaks could be detected. The presence of AgNPs attached to CL nanofibers can be confirmed by diffraction peaks at 2θ values of 38°, 44.5°, 64.5°, and 77.5° indexed to (111), (200), (220), and (311) crystal planes of the standard metal silver pattern, respectively [32,33]. The reduction of silver ions to AgNP formation from the nucleation thus were successfully conducted. As shown in Figure 3B, the TEM image of Ag-CL illustrates the adhesion of AgNPs on to the surface of CL nanofibers with the AgNPs' diameter of 12.8 ± 4.4 nm. In contrast, neat CL nanofibers presented a particle-free morphology. The coating of Ag-CL nanofibers with OEO did not affect the X-ray diffraction (Figure 3A(e),(f)).

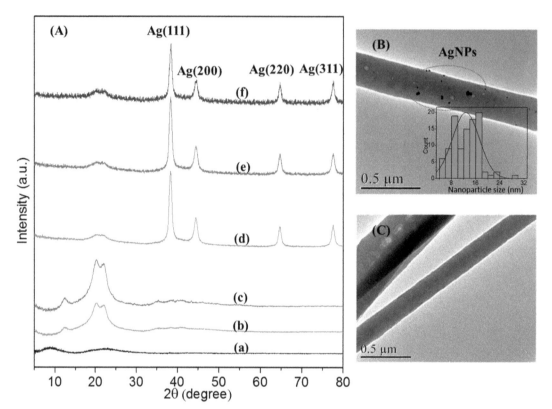

Figure 3. (**A**) XRD spectra of (**a**) CA; (**b**) CL; (**c**) OEO-CL; (**d**) Ag-CL; (**e**) Ag-OEO5%-CL; and (**f**) Ag-OEO10%-CL. (**B**) TEM image of Ag-CL nanofiber. (**C**) TEM image of CL nanofibers.

3.5. EDS Analyses

The elemental analysis was carried out by EDS study (Figure 4). The characteristic results verified that CA and CL nanofibers are composed majorly of carbon and oxygen. For the Ag-CL nanofibers, besides the presence of carbon and oxygen, the detection of silver demonstrated the loading of silver in the samples. Moreover, the nitrogen was not detected in Ag-CL, which verified the complete conversion of $AgNO_3$ to AgNPs during the reducing reaction. The elemental composition is given in Table 2, CA and CL nanofibers were found to be composed of carbon, oxygen, and hydrogen elements. In the case of Ag-CL, the silver element was 3.64%, clarifying the success of AgNPs synthesis.

Table 2. Elemental composition of CA, CL, and Ag-CL nanofibers using EDS analysis.

Sample	Element (at.%)			
	Carbon	Nitrogen	Oxygen	Ag
CA	54.67	-	41.90	-
CL	52.44	-	47.56	-
Ag-CL	44.09	-	52.28	3.64

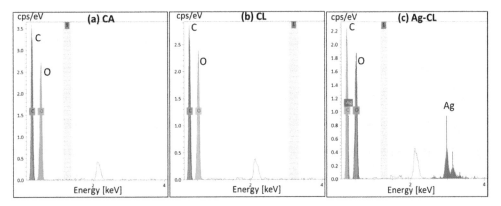

Figure 4. EDS spectra of (**a**) CA, (**b**) CL, and (**c**) Ag-CL nanofibers.

3.6. Mechanical Properties

Table 3 and Figure 5 show the mechanical properties of CA, CL, OEO-CL, Ag-CL, Ag-OEO5%-CL, and Ag-OEO10%-CL. The deacetylation process and mercerization effect increased the strength at break and Young's modulus of as-spun nanofibers from 3.32 ± 0.84 MPa and 144.31 ± 37.74 MPa to 8.51 ± 1.93 MPa and 315.51 ± 45.81 MPa, which are the values of CL nanofibers, respectively. The effect was due to the -OH groups of CL nanofibers, which form hydrogen bonding between inner polymer chains and possibly bind different nanofibers together. The bonding formation led to the void space reduction and tightly packed effect, increasing crystallinity and making the nanofibrous mats stronger and tougher [34]. The incorporation of OEO to CL nanofibers decreases the tensile strength but increases the elongation at break of nanofibrous samples, possibly the integration of essential oil loosens the intermolecular interactions between the polymer chains. Moreover, it could also be explained as the coating effect of OEO freed the linking nodes between different nanofiber strands and helped them move more easily in the tensile tests. However, after treating CL nanofibers with $AgNO_3$ and $NaBH_4$, the Ag-CL nanofibers present poorer mechanical strength and elongation at break. The adverse observation demonstrated the negative effect of AgNPs formation on the interconnective fiber network and the mobility [35,36].

Table 3. Tensile strength, elongation at break, and Young's modulus of CA, CL, OEO-CL, Ag-CL, Ag-OEO5%-CL, and Ag-OEO10%-CL nanofibers.

Sample	Tensile Strength (MPa)	Elongation at Break (%)	Young's Modulus (MPa)
CA	3.32 ± 0.84	17.33 ± 5.79	144.31 ± 37.74
CL	8.51 ± 1.93	3.85 ± 0.98	315.51 ± 45.81
OEO-CL	7.83 ± 3.08	5.23 ± 2.38	254.95 ± 50.16
Ag-CL	6.32 ± 2.81	3.63 ± 1.25	220.27 ± 66.08
Ag-OEO5%-CL	5.91 ± 0.93	4.5 ± 1.91	215.94 ± 30.15
Ag-OEO10%-CL	5.6 ± 1.03	4.96 ± 1.65	204.9 ± 40.48

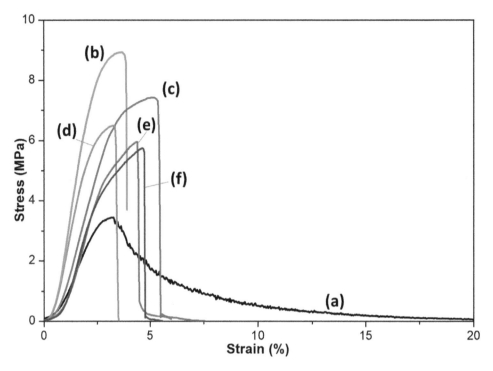

Figure 5. Stress–strain profiles of (**a**) CA, (**b**) CL, (**c**) OEO-CL, (**d**) Ag-CL, (**e**) Ag-OEO5%-CL, and (**f**) Ag-OEO10%-CL.

3.7. Water Absorption and Silver Release Profile

Water absorption was evaluated to quantify the water amount can be absorbed into CA nanofibers, as well as CL nanofibers before and after the incorporation of silver and OEO (Figure 6A). The water holding capacity of CL nanofibers after deacetylation process improved significantly and reached to 1126% compared to CA nanofibers at just 124%. For composite nanofibers, the quantification of water absorption gives insight into the influence of AgNPs and OEO to the hydrophilicity of treated CL nanofibers. The percentage of weight gain of CL nanofibers and Ag-CL nanofibers showed a slight difference, which means the adhesion of AgNPs did not significantly change the hydrophilic properties of the CL nanofibers; AgNPs were considered hydrophobic but the added percentage of silver content proved to be insufficient to change the absorption properties of CL nanofibers. By contrast, OEO addition reduced the water absorption of the CL nanofibers from 1126% to 923% for OEO-CL, and from 1133% to 847% and 776% of Ag-CL, Ag-OEO5%-CL, and Ag-OEO10%-CL, respectively. These findings were in line with other publications [37,38].

The mechanisms of AgNP toxicity are closely related to Ag$^+$ ion release from the polymer matrix into the cell, and controlling the release of silver over time is desirable in wound dressing and other medical applications. Long-term silver release can lead to long-lasting antibacterial activity. The silver release has been reported to correlate to silver content, AgNPs size, water chemistry, and the surface coating of AgNPs. In Figure 6B, AgNP dissolution was studied over a time course of 48 h. Ag-CL presented a speedy release at first, followed by a gradual release curve. The silver discharge of Ag-OEO5%-CL was slower in the first 6 h compared to Ag-CL, and followed by a stable increase, whereas the discharge profile of Ag-OEO10%-CL showed a first-order linear relationship. It is surmised that the phenomenon was ascribable to surface passivation that occurred in the case of Ag-OEO5%-CL and Ag-OEO10%-CL due to the OEO coating effects. The rapid discharge of

Ag at the 6 h time point of Ag-CL was in accordance with other already published articles, which could be explained as the discharge of chemisorbed Ag^+ and the oxidative process with O_2 [39]. The coating effects of OEO cause the slow and sustained release of Ag from the composite nanofibers [40,41].

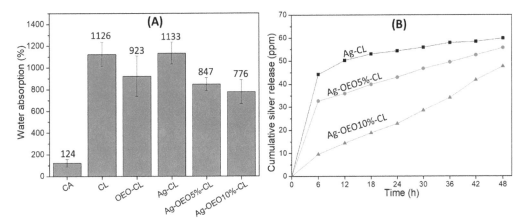

Figure 6. (A) Water absorption of CA, CL, OEO-CL, Ag-CL, Ag-OEO5%-CL, and Ag-OEO10%-CL. (B) Silver release behavior of Ag-CL, Ag-OEO5%-CL, and Ag-OEO10%-CL.

3.8. Antibacterial Activity of Composite CL Nanofibers

The antibacterial effects of CL, OEO-CL, Ag-CL, Ag-OEO5%-CL, and Ag-OEO10%-CL were evaluated against common microorganisms, *E. coli* and *B. subtilis*. The photos of the inhibition zones were taken after incubation time and measured using ImageJ software. Figure 7 and Figure S2 show the diameters of the halo area, generated by the disruption of bacterial growth around the nanofibrous specimens. The CL nanofiber sample or negative control sample presented no antibacterial activities. For all other samples containing AgNPs or OEO, the inhibitions zones were clear and visible, demonstrating the antimicrobial properties against both Gram-positive and Gram-negative bacteria. In the case of *E. coli*, the zone diameters were 7.0 ± 1.0, 6.7 ± 0.8, 7.7 ± 1.6, and 9.8 ± 1.7 mm for OEO-CL, Ag-CL, Ag-OEO5%-CL, and Ag-OEO10%-CL. The combination effects of OEO and AgNPs resulted in improved antibacterial action against *E. coli*, and the sizes of the halo zones were significantly extended from 6.7 ± 0.8 to 9.8 ± 1.7 mm for Ag-CL and Ag-OEO10%-CL, respectively. However, OEO at low content was not so potent against *B. subtilis*—Gram-positive bacteria, when combined with AgNPs, adding OEO to Ag-CL composite nanofibers, the diameters of the halo zones increased from only 7.0 ± 0.5 to 7.4 ± 0.6 for Ag-CL and Ag-OEO5%-CL, respectively. The consistent results were also reported in a previous publication because of the bactericidal effects of OEO on diversities of bacterial cell structures, and OEO is less potent against Gram-positive bacterial strains [42]. OEO coating suppresses the silver release, thus leading to higher OEO concentrations, but the low silver discharge will not effectively extend the sizes of the bacterial inhibition zones against Gram-positive bacteria. The antibacterial properties of AgNPs can be assigned to the reactive oxygen species generation, destroying cell membrane and disturbing the DNA replication and protein synthesis. On the other hand, one of the most important mechanisms of essential oil antibiotics is disturbing the lipid and protein interactions and sabotaging their functions on the basis of their hydrophobicity. The highest inhibition zone was by CL-OEO10% against *E. coli*, which has complex structure of cell wall composed of a thin peptidoglycan layer. In a mixture, EO can interact with other compounds or antibiotic agents, producing antagonistic, additive, indifferent, or synergistic effects [43,44].

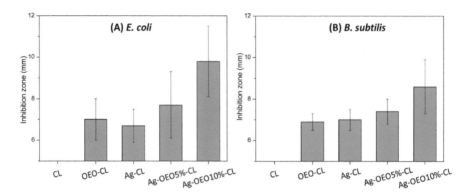

Figure 7. Diameter of inhibition zone with standard deviation of CL, OEO-CL, Ag-CL, Ag-OEO5%-CL, and Ag-OEO10%-CL against (**A**) *E. coli* and (**B**) *B. subtilis*.

4. Conclusions

In recent years, customer interest in environmentally friendly, safe, and natural products has risen. Natural antimicrobial essential oils have antiviral, antimicrobial, and fungicidal benefits; high levels of safety; and a very low impact on human health, with great potential to replace synthetic chemical agents. The drawback can be remedied by combining them with fibrous structures and other metallic antibiotics such as silver, copper, or zinc oxide, which have been reported to offer many favors such as chemical and physical stability. There is some research on essential oils integrated into nanofibers; however, there are still many issues that need to be addressed such as the impacts of oil on mechanical, chemical, and overall performance. Moreover, the synergism when combined with silver is unclear and not yet well reported.

In the present paper, the success of AgNP in situ synthesis and the coating of OEO onto the CL nanofibers was introduced. The good antibacterial properties of AgNPs combined with OEO were demonstrated against *E. coli* and *B. subtilis*. The use of OEO as the coating material represents a sustainable and controllable silver release over the period of 48 h. Interestingly, the antibacterial synergism was great in the case of *E. coli*, whereas OEO content difference did not affect the bactericidal effectiveness against *B. subtilis*. It is concluded that OEO and AgNPs incorporated into CL nanofibers may have the potential as a biomaterial for avoiding bacterial infections in applications such as facial masks, protective clothing, and food packaging. However, more research is needed to further apply EO and AgNPs in commercialized products.

Supplementary Materials: The following are available online at https://www.mdpi.com/article/10.3390/polym14010085/s1, Figure S1. The total ion chromatogram of OEO. Figure S2. Representative photographs of inhibition zone of CL—as the negative control, OEO-CL, Ag-CL, Ag-OEO5%-CL, and Ag-OEO10%-CL.

Author Contributions: D.-N.P.: conceptualization, investigation, formal analysis, and writing original draft; M.Q.K.: formal analysis and writing original draft; V.-C.N.: formal analysis; H.V.-M.: investigation and formal analysis; A.-T.D.: investigation and formal analysis; P.T.T.: investigation and formal analysis; N.-M.N.: formal analysis; V.-T.L.: formal analysis; A.U.: formal analysis; M.K.: formal analysis; I.-S.K.: conceptualization, investigation, and editing. All authors have read and agreed to the published version of the manuscript.

Funding: This research is funded by Hanoi University of Science and Technology (HUST) under grant number T2020-PC-206.

Conflicts of Interest: The authors declare that there are no conflict of interest associated with the work presented.

References

1. Leng, E.; Zhang, Y.; Peng, Y.; Gong, X.; Mao, M.; Li, X.; Yu, Y. In situ structural changes of crystalline and amorphous cellulose during slow pyrolysis at low temperatures. *Fuel* **2018**, *216*, 313–321. [CrossRef]
2. Phan, D.-N.; Lee, H.; Huang, B.; Mukai, Y.; Kim, I.-S. Fabrication of electrospun chitosan/cellulose nanofibers having adsorption.
3. Ahmadzadeh, S.; Nasirpour, A.; Harchegani, M.B.; Hamdami, N.; Keramat, J. Effect of electrohydrodynamic technique as a complementary process for cellulose extraction from bagasse: Crystalline to amorphous transition. *Carbohydr. Polym.* **2018**, *188*, 188–196. [CrossRef]
4. Phan, D.-N.; Khan, M.Q.; Nguyen, N.-T.; Phan, T.-T.; Ullah, A.; Khatri, M.; Kien, N.N.; Kim, I.-S. A review on the fabrication of several carbohydrate polymers into nanofibrous structures using electrospinning for removal of metal ions and dyes. *Carbohydr. Polym.* **2021**, *252*, 117175. [CrossRef]
5. Song, J.; Chen, C.; Yang, Z.; Kuang, Y.; Li, T.; Li, Y.; Huang, H.; Kierzewski, I.; Liu, B.; He, S.; et al. Highly Compressible, Anisotropic Aerogel with Aligned Cellulose Nanofibers. *ACS Nano* **2018**, *12*, 140–147. [CrossRef] [PubMed]
6. Wang, Y.; Guo, Z.; Qian, Y.; Zhang, Z.; Lyu, L.; Wang, Y.; Ye, F. Study on the Electrospinning of Gelatin/Pullulan Composite Nanofibers. *Polymers* **2019**, *11*, 1424. [CrossRef]
7. Barhoum, A.; Pal, K.; Rahier, H.; Uludag, H.; Kim, I.S.; Bechelany, M. Nanofibers as new-generation materials: From spinning and nano-spinning fabrication techniques to emerging applications. *Appl. Mater. Today* **2019**, *17*, 1–35. [CrossRef]
8. Phan, D.-N.; Dorjjugder, N.; Saito, Y.; Taguchi, G.; Ullah, A.; Kharaghani, D.; Kim, I.-S. The synthesis of silver-nanoparticle-anchored electrospun polyacrylonitrile nanofibers and a comparison with as-spun silver/polyacrylonitrile nanocomposite membranes upon antibacterial activity. *Polym. Bull.* **2020**, *77*, 4197–4212. [CrossRef]
9. Xue, J.; Wu, T.; Dai, Y.; Xia, Y. Electrospinning and Electrospun Nanofibers: Methods, Materials, and Applications. *Chem. Rev.* **2019**, *119*, 5298–5415. [CrossRef] [PubMed]
10. Phan, D.-N.; Dorjjugder, N.; Saito, Y.; Khan, M.Q.; Ullah, A.; Bie, X.; Taguchi, G.; Kim, I.-S. Antibacterial mechanisms of various copper species incorporated in polymeric nanofibers against bacteria. *Mater. Today Commun.* **2020**, *25*, 101377. [CrossRef]
11. Phan, D.-N.; Rebia, R.A.; Saito, Y.; Kharaghani, D.; Khatri, M.; Tanaka, T.; Lee, H.; Kim, I.-S. Zinc oxide nanoparticles attached to polyacrylonitrile nanofibers with hinokitiol as gluing agent for synergistic antibacterial activities and effective dye removal. *J. Ind. Eng. Chem.* **2020**, *85*, 258–268. [CrossRef]
12. Gaminian, H.; Montazer, M. Decorating silver nanoparticles on electrospun cellulose nanofibers through a facile method by dopamine and ultraviolet irradiation. *Cellulose* **2017**, *24*, 3179–3190. [CrossRef]
13. Yin, I.X.; Zhang, J.; Zhao, I.S.; Mei, M.L.; Li, Q.; Chu, C.H. The Antibacterial Mechanism of Silver Nanoparticles and Its Application in Dentistry. *Int. J. Nanomed.* **2020**, *15*, 2555–2562. [CrossRef]
14. Panáček, A.; Kvítek, L.; Smékalová, M.; Večeřová, R.; Kolář, M.; Röderová, M.; Dyčka, F.; Šebela, M.; Prucek, R.; Tomanec, O.; et al. Bacterial resistance to silver nanoparticles and how to overcome it. *Nat. Nanotechnol.* **2018**, *13*, 65–71. [CrossRef]
15. Burduşel, A.-C.; Gherasim, O.; Grumezescu, A.M.; Mogoantă, L.; Ficai, A.; Andronescu, E. Biomedical Applications of Silver Nanoparticles: An Up-to-Date Overview. *Nanomaterials* **2018**, *8*, 681. [CrossRef]
16. Phan, D.-N.; Dorjjugder, N.; Saito, Y.; Taguchi, G.; Lee, H.; Lee, J.S.; Kim, I.-S. The mechanistic actions of different silver species at the surfaces of polyacrylonitrile nanofibers regarding antibacterial activities. *Mater. Today Commun.* **2019**, *21*, 100622. [CrossRef]
17. Castangia, I.; Marongiu, F.; Manca, M.L.; Pompei, R.; Angius, F.; Ardu, A.; Fadda, A.M.; Manconi, M.; Ennas, G. Combination of grape extract-silver nanoparticles and liposomes: A totally green approach. *Eur. J. Pharm. Sci.* **2017**, *97*, 62–69. [CrossRef] [PubMed]
18. Giunti, G.; Palermo, D.; Laudani, F.; Algeri, G.M.; Campolo, O.; Palmeri, V. Repellence and acute toxicity of a nano-emulsion of sweet orange essential oil toward two major stored grain insect pests. *Ind. Crop. Prod.* **2019**, *142*, 111869. [CrossRef]
19. Gavahian, M.; Chu, Y.; Khaneghah, A.M. Recent advances in orange oil extraction: An opportunity for the valorisation of orange peel waste a review. *Int. J. Food Sci. Technol.* **2018**, *54*, 925–932. [CrossRef]
20. Evangelho, J.A.D.; Dannenberg, G.D.S.; Biduski, B.; el Halal, S.L.M.; Kringel, D.H.; Gularte, M.A.; Fiorentini, A.M.; Zavareze, E.D.R. Antibacterial activity, optical, mechanical, and barrier properties of corn starch films containing orange essential oil. *Carbohydr. Polym.* **2019**, *222*, 114981. [CrossRef] [PubMed]
21. Felix de Andrade, M.; Diego de Lima Silva, I.; Alves da Silva, G.; David Cavalcante, P.V.; Thayse da Silva, F.; Bastos de Almeida, Y.M.; Vinhas, G.M.; Hecker de Carvalho, L. A study of poly (butylene adipate-co-terephthalate)/orange essential oil films for application in active antimicrobial packaging. *LWT* **2020**, *125*, 109148. [CrossRef]
22. Phan, D.-N.; Dorjjugder, N.; Khan, M.Q.; Saito, Y.; Taguchi, G.; Lee, H.; Mukai, Y.; Kim, I.-S. Synthesis and attachment of silver and copper nanoparticles on cellulose nanofibers and comparative antibacterial study. *Cellulose* **2019**, *26*, 6629–6640. [CrossRef]
23. Golmohammadi, M.; Borghei, A.; Zenouzi, A.; Ashrafi, N.; Taherzadeh, M. Optimization of essential oil extraction from orange peels using steam explosion. *Heliyon* **2018**, *4*, e00893. [CrossRef]
24. Yang, C.; Chen, H.; Chen, H.; Zhong, B.; Luo, X.; Chun, J. Antioxidant and Anticancer Activities of Essential Oil from Gannan Navel Orange Peel. *Molecules* **2017**, *22*, 1391. [CrossRef] [PubMed]
25. Liu, K.; Deng, W.; Hu, W.; Cao, S.; Zhong, B.; Chun, J. Extraction of 'Gannanzao' Orange Peel Essential Oil by Response Surface Methodology and its Effect on Cancer Cell Proliferation and Migration. *Molecules* **2019**, *24*, 499. [CrossRef]

26. Hassan, M.Z.; Roslan, S.A.; Sapuan, S.M.; Rasid, Z.A.; Mohd Nor, A.F.; Md Daud, M.Y.; Dolah, R.; Mohamed Yusoff, M.Z. Mercerization Optimization of Bamboo (*Bambusa vulgaris*) Fiber-Reinforced Epoxy Composite Structures Using a Box–Behnken Design. *Polymers* **2020**, *12*, 1367. [CrossRef] [PubMed]

27. Nakagaito, A.N.; Yano, H. Toughness enhancement of cellulose nanocomposites by alkali treatment of the reinforcing cellulose nanofibers. *Cellulose* **2008**, *15*, 323–331. [CrossRef]

28. Lu, Y.; Wang, H.; Lu, Y. An architectural exfoliated-graphene carbon aerogel with superhydrophobicity and efficient selectivity. *Mater. Des.* **2019**, *184*, 108134.

29. Ahmed, F.; Arbab, A.A.; Jatoi, A.W.; Khatri, M.; Memon, N.; Khatri, Z.; Kim, I.S. Ultrasonic-assisted deacetylation of cellulose acetate nanofibers: A rapid method to produce cellulose nanofibers. *Ultrason. Sonochem.* **2017**, *36*, 319–325. [CrossRef]

30. Lu, Y.; Ye, G.; She, X.; Wang, S.; Yang, D.; Yin, Y. Sustainable Route for Molecularly Thin Cellulose Nanoribbons and Derived Nitrogen-Doped Carbon Electrocatalysts. *ACS Sustain. Chem. Eng.* **2017**, *5*, 8729–8737. [CrossRef]

31. Cebi, N.; Taylan, O.; Abusurrah, M.; Sagdic, O. Detection of Orange Essential Oil, Isopropyl Myristate, and Benzyl Alcohol in Lemon Essential Oil by FTIR Spectroscopy Combined with Chemometrics. *Foods* **2020**, *10*, 27. [CrossRef] [PubMed]

32. Li, J.; Ma, R.; Wu, Z.; He, S.; Chen, Y.; Bai, R.; Wang, J. Visible-Light-Driven Ag-Modified TiO$_2$ Thin Films Anchored on Bamboo Material with Antifungal Memory Activity against *Aspergillus niger*. *J. Fungi* **2021**, *7*, 592. [CrossRef] [PubMed]

33. Li, J.; Ma, R.; Lu, Y.; Wu, Z.; Su, M.; Jin, K.; Qin, D.; Zhang, R.; Bai, R.; He, S.; et al. A gravity-driven high-flux catalytic filter prepared using a naturally three-dimensional porous rattan biotemplate decorated with Ag nanoparticles. *Green Chem.* **2020**, *22*, 6846–6854. [CrossRef]

34. Maroufi, L.Y.; Ghorbani, M.; Mohammadi, M.; Pezeshki, A. Improvement of the physico-mechanical properties of antibacterial electrospun poly lactic acid nanofibers by incorporation of guar gum and thyme essential oil. *Colloids Surf. A Physicochem. Eng. Asp.* **2021**, *622*, 126659. [CrossRef]

35. Alizadeh-Sani, M.; Khezerlou, A.; Ehsani, A. Fabrication and characterization of the bionanocomposite film based on whey protein biopolymer loaded with TiO2 nanoparticles, cellulose nanofibers and rosemary essential oil. *Ind. Crop. Prod.* **2018**, *124*, 300–315. [CrossRef]

36. Jahed, E.; Khaledabad, M.A.; Bari, M.R.; Almasi, H. Effect of cellulose and lignocellulose nanofibers on the properties of Origanum vulgare ssp. gracile essential oil-loaded chitosan films. *React. Funct. Polym.* **2017**, *117*, 70–80. [CrossRef]

37. Ullah, A.; Saito, Y.; Ullah, S.; Haider, K.; Nawaz, H.; Duy-Nam, P.; Kharaghani, D.; Kim, I.S. Bioactive Sambong oil-loaded electrospun cellulose acetate nanofibers: Preparation, characterization, and in-vitro biocompatibility. *Int. J. Biol. Macromol.* **2021**, *166*, 1009–1021. [CrossRef]

38. Sharma, A.; Mandal, T.; Goswami, S. Fabrication of cellulose acetate nanocomposite films with lignocelluosic nanofiber filler for superior effect on thermal, mechanical and optical properties. *Nano-Struct. Nano-Objects* **2021**, *25*, 100642. [CrossRef]

39. Dobias, J.; Bernier-Latmani, R. Silver Release from Silver Nanoparticles in Natural Waters. *Environ. Sci. Technol.* **2013**, *47*, 4140–4146. [CrossRef]

40. Boccalon, E.; Pica, M.; Romani, A.; Casciola, M.; Sterflinger, K.; Pietrella, D.; Nocchetti, M. Facile preparation of organic-inorganic hydrogels containing silver or essential oil with antimicrobial effects. *Appl. Clay Sci.* **2020**, *190*, 105567. [CrossRef]

41. Wu, Z.; Zhou, W.; Pang, C.; Deng, W.; Xu, C.; Wang, X. Multifunctional chitosan-based coating with liposomes containing laurel essential oils and nanosilver for pork preservation. *Food Chem.* **2019**, *295*, 16–25. [CrossRef]

42. Farahmandfar, R.; Tirgarian, B.; Dehghan, B.; Nemati, A. Comparison of different drying methods on bitter orange (*Citrus aurantium* L.) peel waste: Changes in physical (density and color) and essential oil (yield, composition, antioxidant and antibacterial) properties of powders. *J. Food Meas. Charact.* **2019**, *14*, 862–875. [CrossRef]

43. Amjadi, S.; Almasi, H.; Ghorbani, M.; Ramazani, S. Reinforced ZnONPs/ rosemary essential oil-incorporated zein electrospun nanofibers by κ-carrageenan. *Carbohydr. Polym.* **2020**, *232*, 115800. [CrossRef] [PubMed]

44. Vafania, B.; Fathi, M.; Soleimanian-Zad, S. Nanoencapsulation of thyme essential oil in chitosan-gelatin nanofibers by nozzle-less electrospinning and their application to reduce nitrite in sausages. *Food Bioprod. Process.* **2019**, *116*, 240–248. [CrossRef]

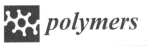 *polymers*

Article

A Flexible, Fireproof, Composite Polymer Electrolyte Reinforced by Electrospun Polyimide for Room-Temperature Solid-State Batteries

Boheng Yuan [1,†], Bin Zhao [1,†], Zhi Cong [1], Zhi Cheng [1], Qi Wang [1], Yafei Lu [1] and Xiaogang Han [1,2,*]

1 State Key Laboratory of Electrical Insulation and Power Equipment, School of Electrical Engineering, Xi'an Jiaotong University, Xi'an 710049, China; yuan980808@stu.xjtu.edu.cn (B.Y.); zhaobin87@xjtu.edu.cn (B.Z.); jycz1997@stu.xjtu.edu.cn (Z.C.); chengzhi@stu.xjtu.edu.cn (Z.C.); wq4116005096@stu.xjtu.edu.cn (Q.W.); luyafei0126@stu.xjtu.edu.cn (Y.L.)
2 Key Laboratory of Smart Grid of Shanxi Province, School of Electrical Engineering, Xi'an Jiaotong University, Xi'an 710049, China
* Correspondence: xiaogang.han@xjtu.edu.cn
† Bin Zhao and Boheng Yuan contributed equally to this work.

Abstract: Solid-state batteries (SSBs) have attracted considerable attention for high-energy-density and high-safety energy storage devices. Many efforts have focused on the thin solid-state-electrolyte (SSE) films with high room-temperature ionic conductivity, flexibility, and mechanical strength. Here, we report a composite polymer electrolyte (CPE) reinforced by electrospun PI nanofiber film, combining with succinonitrile-based solid composite electrolyte. In situ photo-polymerization method is used for the preparation of the CPE. This CPE, with a thickness around 32.5 μm, shows a high ionic conductivity of 2.64×10^{-4} S cm^{-1} at room temperature. It is also fireproof and mechanically strong, showing great promise for an SSB device with high energy density and high safety.

Keywords: electrospinning; polyimide; solid state batteries; composite polymer electrolyte; photo polymerization; fireproof

Citation: Yuan, B.; Zhao, B.; Cong, Z.; Cheng, Z.; Wang, Q.; Lu, Y.; Han, X. A Flexible, Fireproof, Composite Polymer Electrolyte Reinforced by Electrospun Polyimide for Room-Temperature Solid-State Batteries. *Polymers* **2021**, *13*, 3622. https://doi.org/10.3390/polym13213622

Academic Editor: Ick-Soo Kim

Received: 21 September 2021
Accepted: 12 October 2021
Published: 20 October 2021

Publisher's Note: MDPI stays neutral with regard to jurisdictional claims in published maps and institutional affiliations.

1. Introduction

Decarbonization is the global trend, which has led to considerable attention on the research of battery-energy storage systems enabling high-efficiency utilization of renewable solar and wind energy. Li-ion batteries have become dominant energy storage device in portable electronic devices and electric vehicles [1–3]. However, energy density and safety of the current state-of-the-art Li-ion batteries still do not meet the requirements. For this concern, Li-metal batteries with metallic Li anode have been intensively studied [4–7]. The main problems of liquid Li-metal batteries include side reaction between electrolyte and lithium metal and the growth of lithium dendrites, which can be addressed by solid-state Li-metal batteries that are considered as next generation batteries-technology [8–11].

In solid-state batteries (SSBs), solid-state electrolytes (SSEs) lie at the heart. SSEs can be summarized in three categories: inorganic solid electrolytes (ISEs) [12–14], solid polymer electrolytes (SPEs) [15–19], and composite polymer electrolytes (CPEs) [20–22]. For an ideal SSE, it should have high ionic conductivity, strong mechanical strength, good flexibility, wide electrochemical stability, nonflammability, etc. ISEs own the highest ionic conductivity among all types of SSEs, but they are hard to fabricate thin-film electrolyte with high flexibility and mechanical strength, especially for the oxide-based solid electrolyte. Compared with ISEs, SPEs, composed of uniform mixtures of polymer host and Li salt, have attracted much attention for their lightweight, high flexibility, and ease of processing. The main issues that hinder the application of SPEs are low room-temperature conductivity, and weak mechanical strength to suppress Li dendrite. To circumvent these problems, CPEs, combining SPEs with nanoparticle fillers, plasticizers, or 3D robust host, are considered more suitable as solid electrolyte [23–27]. CPE films with high room-temperature Li-ion conductivity and strong mechanical strength is the key for high performance SSBs. To achieve this goal, we studied two strategies to address the ionic conductivity and mechanical strength issues. (1) Succinonitrile (SN) plastic crystal is a non-ionic plastic material, with molecules ordered into a crystalline lattice, meanwhile exhibiting a fluctuational degree of freedom. In recent years, SN have been widely introduced into CPEs system as versatile additive to enhance the room-temperature Li-ion conductivity [28–33]. (2) In another side, to enhance the mechanical strength, many kinds of 3D frameworks have been studied as supporting matrix, including organic and inorganic networks [24–26]. Polyimide (PI) is an insulating material that has excellent thermal (>300 °C) and chemical stabilities. There are already many studies using PI as functional separator, as well as supporting matrix for SPEs [34–39]. Besides, PI owns a novel property of fireproof, which is very important for the safety of batteries.

In this work, we proposed a type of CPEs with high room-temperature Li-ion conductivity and good mechanical flexibility via UV polymerization method. In this CPE, SN is added as plasticizer and PI nano-fiber film is used as skeleton. The CPE film shows a very good flexibility, and a high room-temperature ionic conductivity of 2.64×10^{-4} S cm^{-1}. A ASSLB cell assembled with LiCoO$_2$ (LCO) cathode shows an initial discharge capacity of 131.1 mAh g^{-1}, and 83.3% retention after 120 cycles. The CPE film also shows a novel fireproof property that enables promising safe ASSLBs.

2. Materials and Methods

2.1. Preparation of Polyimide (PI) Fiber

The PI fiber film was prepared via electrospinning deposition followed by an imidization process. In details, 4,4′-oxybisbenzenamine (ODA, ≥98%, Aladdin) was added to dimethylacetamide (DMAC, 99.9%, Sinopharm Chemical Reagent Co., Ltd., Shanghai, China) under magnetic stirring. Then, 1,2,4,5-Benzenetetracarboxylic anhydride (PMDA, ≥99%, Aladdin, Shanghai, China) was added after ODA completely dissolved, where the weight ratio of ODA to PMDA was 1:1, to get polyamide acid (PAA) solution after 12 h stirring. The concentration of PAA solution is 15 wt.%.

For electrospinning, the voltage was 20 kV, the distance between needle and collecting plate (roller collector 42BYGH47-401A, Alibaba Group, Hangzhou, China) was 18 cm and the propulsion speed was 1 ml h^{-1}. Then PAA nanofiber membranes were dehydrated and cyclized at 100 °C for 2 h and 200 °C for 2 h and 300 °C for 2 h, respectively, in a muffle furnace to get the final product of PI nanofiber film.

2.2. Preparation of PI–CPE

LiTFSI (99%, Aladdin) and SN (99%, Macklin, Shanghai, China) were dissolved in PEGDA (~700, Macklin) in Ar filled glovebox with a mass ratio of 4:4:2 followed by magnetical stirring for 4 h, then 1 wt.% of photoinitiator CIBA (IRGACURE 819) (with respect to PEGDA) were added to obtain homogeneous precursor solution. To stabilize the interface between electrolyte and Li metal, 1 wt.% of LiNO$_3$ (99.99%, Aladdin, with respect to total weight) was added to the precursor solution. The as-prepared precursor solution was dipped onto PI nanofiber film and allowed to stand for 5 min to let solution filtrated into the PI film. The obtained composite membrane was sandwiched between two glass slides and a UV-cured with UVLED machine (UVB 365 nm, XM210, Aventk, Shanghai, China). The prepared CPE film is named PI–CPE.

2.3. Characterization of Samples

The morphology of samples was characterized by a scanning electron microscopy (SEM, Phenom Pro X, Phenom Scientific, Shanghai, China). Fourier transform infrared spectrometer (ATR-FT-IR, IN10+IZ10, Nicolet, Madison, WI, USA) over the range from 400 to 4000 cm^{-1}. Mechanical strength of PI-CPE was evaluated by a dynamic mechanical thermal analyzer (DMA, Q800, TA instruments, New Castle, DE, USA). The phase transition of materials during heating were investigated by differential scanning calorimeter (DSC, DSC822E, METTLER TOLEDO, Shanghai, China) performed from −100 °C to 150 °C.

2.4. Electrochemical Characterization

The PI–CPE film was evaluated as a solid electrolyte in stainless steel (SS) | | SS, SS | | Li, Li | | Li, and Li | | LiCoO$_2$ (LCO) cell configurations. The ionic conductivity of PI-CPE was measured by Bio-Logic SP-500 with impedance spectra method of SS | | SS cell over a frequency range between 7 MHz and 1 Hz at a temperature range from room temperature (RT) to 80 °C. The electrochemical window of the PI–CPE was tested by electrochemical floating analysis and linear sweep voltammetry (LSV) at a sweep rate of 1 mV s^{-1} from 0 to 6 V with carbon coated Al foil as the working electrode and Li metal as the counter electrode (C–Al | | Li cell). The lithium ion transference number (t_{Li^+}) of PI–CPE was evaluated by combining alternating current impedance and direct-current (DC) polarization with a DC voltage of 10mV using a Li | | Li cell. The stability of PI–CPE with Li metal was carried out by Li plating/stripping cycles at a current density of 0.1 mA cm^{-2} with Li | | Li symmetric cell on a LAND CT2001A testing system.

2.5. Evaluation of Solid-State Batteries

Cathode slurry was prepared by grinding LCO powder, carbon black, and polyvinylidene difluoride powder (PVDF) with mass fractions of 80%, 10%, and 10%, respectively, followed by the addition of methylpyrrolidone (NMP) solvent. The cathode slurry was then casted onto aluminum film, and dried in a vacuum oven at 80 °C for 12 h. Precursor electrolyte solution was dipped onto cathode film and kept for 2 h; then, the well-infiltrated cathode and PI film infiltrated with precursor solution were stacked together and in situ photo-polymerization. Coin cells were assembled with Li foil as anode and charging/discharging tests of batteries were performed on a LAND CT2001A testing system at room temperature.

3. Results

3.1. Structural Characterization of PI–CPE

The schematic in Figure 1 shows the preparation progress of PI–CPE. PI nanofiber obtained via electrospinning is firstly filled with precursor solution. Notably, there is no additional solvent existing in the precursor, and it is a solvent-free process. Under UV irradiation, PEDGA will be polymerized, and the precursor solution changes from liquid to solid. Figure S1 shows the photo image of the as-prepared PI film. The morphology of PI nanofiber is characterized by SEM, shown in Figure 2. The PI nanofibers all have a smooth surface, and much space formed the inner of the PI film caused by the stacking of PI fibers. So, PI-nanofiber film can serve as 3D framework to support the solid electrolyte. Figure S2 shows the photo image of PI–CPE film. The PI-CPE film has a very flexible nature. When put on a glass tube, PI–CPE can naturally droop due to the gravity. To further verify the flexibility of PI–CPE, a simple twisting test is performed (Figure 3). When PI—CPE recovers from twisting, it can regain its original morphology with a smooth surface, and no scratch appears on the surface.

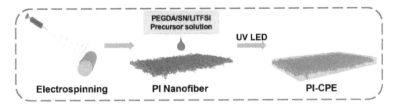

Figure 1. Schematic of the preparation process.

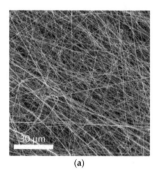 

(a)   (b)

Figure 2. (a,b) are SEM images of PI nanofiber at different magnifications.

Figure 3. Twisting test of PI–CPE film.

The solidification of precursor solution under UV irradiation is contributed by the crosslinking of PEGDA molecule. To verify the reaction, FTIR of PI–CPE, precursor solution, and pure SN/LiTFSI without PEGDA are performed, and results are shown in Figure 4. Without PEGDA, spectra of SN/LiTFSI shows a flat line, and no peak can be seen at the wavenumber range from 1600 to 1800 cm^{-1}. Compared with precursor solution, after UV polymerization, the characteristic peak of C=C at 1617 cm^{-1} of PI–CPE disappears, and the adsorption peak of C=O stretching at 1713 cm^{-1} blue shifts to 1731 cm^{-1}, which confirms the breaking of C=C bonds and involvement of PEG back-bones. So, the PI–CPE has been well UV cured for a free-standing solid electrolyte. The peak is present at 1640 cm^{-1} with respect to the OH bending mode of water due to the absorption of moisture when exposed to air.

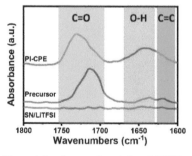

Figure 4. Fourier transform infrared (FTIR) spectra of PI–CPE.

After polymerization, the morphology of as the prepared PI–CPE is characterized with SEM. Figure 5a,b is the top-surface SEM images of PI–CPE, showing a dense structure and no PI fiber or holes can be found at the surface. To make sure the dense structure inside the PI–CPE, cross-section SEM images were also collected, as shown in Figure 6a,b. The inner of PI–CPE is fully filled with CPE with no pore structure. So, the dense structure of PI–CPE is further confirmed. The thickness of PI–CPE is about 32.5 μm. Figure S3 demonstrates the stress-strain curves of PI–CPE film. The tensile strength of PI–CPE can reach as high as 5.2 MPa with an elongation-at-break of 26%. A good mechanical strength can enable PI–CPE, suppressing the Li dendrite growth.

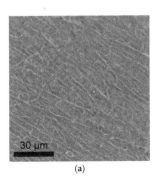
(a)

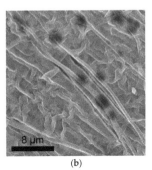
(b)

Figure 5. (a) and (b) are top-surface SEM images of PI–CPE at different magnifications.

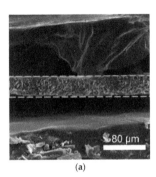
(a)

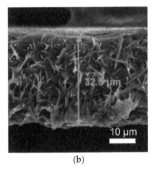
(b)

Figure 6. (a) and (b) are cross-sectional SEM images of PI–CPE at different magnifications.

3.2. Electrochemical Performance

Figure S4 shows the DSC curve of PI–CPE, and the low glass transition temperature of about $-38.5\,°C$ can benefit the room-temperature ionic conductivity. The ionic conductivity of PI–CPE is measured by SS | | SS cell at a temperature range from RT to 80 °C, and EIS plots are shown in Figure 7a. The Li-ion conductivity of PI–CPE is as high as 2.64×10^{-4} S cm^{-1} at room temperature (25 °C). This high ionic conductivity combine with the relatively small thickness (32.5 μm) make PI–CPE qualified for practical application. The ionic conductivities of PI–CPE are 3.18×10^{-4} S cm^{-1}, 4.54×10^{-4} S cm^{-1}, 6.37×10^{-4} S cm^{-1}, 8.49×10^{-4} S cm^{-1}, 1.06×10^{-3} S cm^{-1}, and 1.23×10^{-3} S cm^{-1} at 30 °C, 40 °C, 50 °C, 60 °C, 70 °C, and 80 °C, respectively. Figure 7b shows the Arrhenius plot, and the calculated activation energy E_a of PI–CPE by Arrhenius equation is 0.11 eV. For a solid electrolyte film, not only ionic conductivity but also Li-ion transference number (t_{Li^+}) is a key factor that significant influent the cycle performance of SSBs. Too much anion transfer will cause polarization. DC polarization and alternating current impedance technology are employed together to measure t_{Li^+} of PI–CPE (Figure 8a). The t_{Li^+} of PI–CPE calculated by Bruce–Vincent–Evans equation is 0.76. It is much higher than the PEO-based polymer electrolyte, which is commonly around 0.2. Electrochemical window is another important parameter for SSEs. A wide electrochemical window can make SSEs compatible with high-voltage cathode, thus giving a higher energy density. Figure 8b shows the electrochemical floating analysis and LSV curve (inset) of PI–CPE, carbon-coated Al foil can enlarge the contact area between current collector and PI–CPE. Electrochemical floating analysis provides a stringent test of the oxidative stability of PI–CPE, and the electrochemical stable window is around 4.6 V, which is consistent with LSV test (Figure 8b).

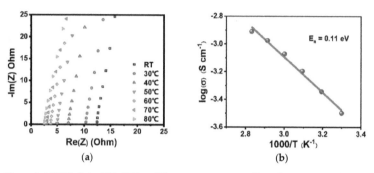

Figure 7. (a) EIS plots of PI–CPE at different temperatures; (b) Arrhenius plots of PI–CPE.

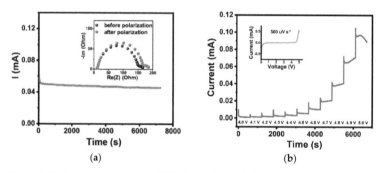

Figure 8. (a) Current-time curve and EIS plots before and after polarization for PI–CPE with Li | | Li cell. (b) Electrochemical floating analysis with C–Al | PI–CPE | Li cell; inset is the LSV curve at a scan rate of 500 uV s⁻¹.

The Li | | Li symmetric cell is used to test whether this PI–CPE is compatible with the Li metal anode. Cycling test was carried out under a density of 0.1 mA cm² and 1 h for each cycle at room temperature in Figure 9. The symmetric cell shows stable Li plating and stripping process. After cycling for 400 h, the overpotential only slightly increases from 18 mV to 24 mV. The good compatibility of PI–CPE with Li anode make it qualified for the SSBs. LCO | PI–CPE | Li coin cells are made to test the performance of PI–CPE film as SSE. LCO cathode is filled with the precursor solution same as PI–CPE and cured under a UV lamp. The LCO | PI–CPE | Li coin cell was cycled at room temperature. Figure 10a shows the voltage profile with a rate of 0.2 C. The initial charge and discharge specific capacity are 136.8 mAh g⁻¹ and 131.1 mAh g⁻¹, and the first-cycle coulombic efficiency is 95.8%. After 100 cycles, the over potential at the 100th cycle is no big difference with the 1st cycle, which shows good cycling stability. The LCO | PI–CPE | Li cell presents excellent cycling performance with 83.3% discharge capacity retention after 120 cycles at 0.2 C under room temperature. The high-voltage-compatible is also demonstrated in this test because of the use of LCO high voltage cathode.

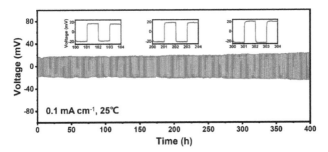

Figure 9. Cycling performance of Li | PI–CPE | Li symmetric cell.

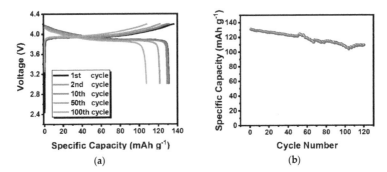

Figure 10. Voltage profile (**a**) and cycling performance (**b**) of PI–CPE at 0.2 C; cycling test was performed at room temperature.

3.3. Thermal and Fireproof Property

Thermal stability is a very important parameter, which will reduce the risk of thermal runaway caused by short circuit. Commercial PP separator is used as comparison. PP and PI–CPE were heated from room temperature to 80 °C and 140 °C (Figure 11). From the pictures we can see that, when heated to 140 °C, PP shrinks to a point, and PI–CPE can still maintain its initial structure. The excellent stability of PI–CPE is mainly because the supporting matrix of PI fiber film. To test the fireproof property of PI–CPE, a flame test was also adopted. As shown in Figure 12, when exposed to the flame for 2 s, PI–CPE did not catch fire, showing a flame retardant property. Firstly, PI is highly thermally stable. Besides, SN will release CO, CO_2, NO gas, etc., under high temperature, which can serve as fire extinguishing agent. So, the as-prepared PI–CPE is fireproof that is promising for high-safe SSBs.

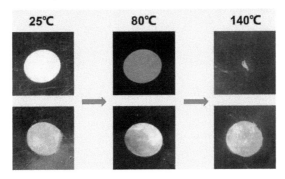

Figure 11. Comparison of thermal stability between commercial PP separator and PI–CPE.

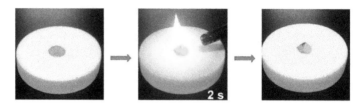

Figure 12. Flame test of PI–CPE.

4. Conclusions

We have proposed a flexible and fireproof polymer-polymer CPE thin film. Using SN as a plasticizer, the as-prepared PI–CPE shows a high room-temperature ionic conductivity of 2.64×10^{-4} S cm^{-1}, which enables it suitable for SSBs running at ambient temperature. Due to the supporting of electrospun PI fiber, the PI–CPE has a high mechanical strength, giving an excellent ability suppressing Li dendrite growth. Moreover, the PI–CPE shows good thermal stability and is fireproof for SSBs with high safety. The Li | PI–CPE | Li symmetric cell can stable cycles more than 400 h, and the LCO | PI–CPE | Li full cell give an initial discharge specific capacity of 131.1 mAh g^{-1} with 83.3% discharge capacity retention after 120 cycles at 0.2 C under room temperature. This CPE is very promising for practical application of SSBs.

Supplementary Materials: The following are available online at https://www.mdpi.com/article/10.3390/polym13213622/s1, Figure S1: Photograph of PI film, Figure S2: Photograph of PI–CPE film, Figure S3: Strain–stress curves of PI–CPE film, Figure S4: DSC curve of PI–CPE.

Author Contributions: Conceptualization, B.Z.; Data curation, X.H.; Formal analysis, B.Z. and Z.C. (Zhi Cong); Funding acquisition, B.Z. and X.H.; Investigation, B.Y.; Methodology, B.Z.; Project administration, X.H.; Resources, X.H.; Software, Z.C. (Zhi Cheng); Supervision, X.H.; Validation, B.Z., B.Y., Q.W. and Y.L.; Visualization, B.Z.; Writing–original draft, B.Y.; Writing–review & editing, B.Z. All authors have read and agreed to the published version of the manuscript.

Funding: This research was funded by National Key R&D Program of China (Grant No. 2018YFB0104300), National Natural Science Foundation of China (Grant No. 51707151), National Natural Science Foundation of China (Grant No. 51772241).

Institutional Review Board Statement: Not applicable.

Informed Consent Statement: Not applicable.

Data Availability Statement: Not applicable.

Acknowledgments: X. Han would like to thank the Independent Research Project of State Key Laboratory of Electrical Insulation and Power Equipment for the financial support. B. Zhao acknowledge the State Key Laboratory of Electrical Insulation and Power Equipment for financial support.

Conflicts of Interest: The authors declare no conflict of interest.

References

1. Armand, M.; Tarascon, J.M. Building better batteries. *Nature* **2008**, *451*, 652–657. [CrossRef] [PubMed]
2. Goodenough, J.B.; Park, K.-S. The Li-Ion rechargeable battery: A perspective. *J. Am. Chem. Soc.* **2013**, *135*, 1167–1176. [CrossRef] [PubMed]
3. Harper, G.; Sommerville, R.; Kendrick, E.; Driscoll, L.; Slater, P.; Stolkin, R.; Walton, A.; Christensen, P.; Heidrich, O.; Lambert, S.; et al. Recycling lithium-ion batteries from electric vehicles. *Nature* **2019**, *575*, 75–86. [CrossRef] [PubMed]
4. He, X.; Bresser, D.; Passerini, S.; Baakes, F.; Krewer, U.; Lopez, J.; Mallia, C.T.; Shao-Horn, Y.; Cekic-Laskovic, I.; Wiemers-Meyer, S.; et al. The passivity of lithium electrodes in liquid electrolytes for secondary batteries. *Nat. Rev. Mater.* **2021**. [CrossRef]
5. Lin, D.; Liu, Y.; Cui, Y. Reviving the lithium metal anode for high-energy batteries. *Nat. Nanotechnol.* **2017**, *12*, 194–206. [CrossRef]
6. Xiao, J.; Li, Q.; Bi, Y.; Cai, M.; Dunn, B.; Glossmann, T.; Liu, J.; Osaka, T.; Sugiura, R.; Wu, B.; et al. Understanding and applying coulombic efficiency in lithium metal batteries. *Nat. Energy* **2020**, *5*, 561–568. [CrossRef]
7. Zou, P.; Sui, Y.; Zhan, H.; Wang, C.; Xin, H.L.; Cheng, H.-M.; Kang, F.; Yang, C. Polymorph evolution mechanisms and regulation strategies of lithium metal anode under multiphysical fields. *Chem. Rev.* **2021**, *121*, 5986–6056. [CrossRef]

8. Fan, L.-Z.; He, H.; Nan, C.-W. Tailoring inorganic-polymer composites for the mass production of solid-state batteries. *Nat. Rev. Mater.* **2021**. [CrossRef]
9. Liu, J.; Yuan, H.; Liu, H.; Zhao, C.-Z.; Lu, Y.; Cheng, X.-B.; Huang, J.-Q.; Zhang, Q. Unlocking the failure mechanism of solid state lithium metal batteries. *Adv. Energy Mater.* **2021**. [CrossRef]
10. Sun, N.; Liu, Q.; Cao, Y.; Lou, S.; Ge, M.; Xiao, X.; Lee, W.-K.; Gao, Y.; Yin, G.; Wang, J.; et al. Anisotropically electrochemical-mechanical evolution in solid-State batteries and interfacial tailored strategy. *Angew. Chem. Int. Ed.* **2019**, *58*, 18647–18653. [CrossRef]
11. Wang, C.; Fu, K.; Kammampata, S.P.; McOwen, D.W.; Samson, A.J.; Zhang, L.; Hitz, G.T.; Nolan, A.M.; Wachsman, E.D.; Mo, Y.; et al. Garnet-type solid-state electrolytes: Materials, interfaces, and batteries. *Chem. Rev.* **2020**, *120*, 4257–4300. [CrossRef]
12. Famprikis, T.; Canepa, P.; Dawson, J.A.; Islam, M.S.; Masquelier, C. Fundamentals of inorganic solid-state electrolytes for batteries. *Nat. Mater.* **2019**, *18*, 1278–1291. [CrossRef]
13. Han, X.; Gong, Y.; Fu, K.; He, X.; Hitz, G.T.; Dai, J.; Pearse, A.; Liu, B.; Wang, H.; Rublo, G.; et al. Negating interfacial impedance in garnet-based solid-state Li metal batteries. *Nat. Mater.* **2017**, *16*, 572. [CrossRef] [PubMed]
14. Shen, F.; Guo, W.; Zeng, D.; Sun, Z.; Gao, J.; Li, J.; Zhao, B.; He, B.; Han, X. A simple and highly efficient method toward high-density garnet-type LLZTO solid-state electrolyte. *ACS Appl. Mater. Interfaces* **2020**, *12*, 30313–30319. [CrossRef] [PubMed]
15. Zhang, H.; Zhang, J.; Ma, J.; Xu, G.; Dong, T.; Cui, G. Polymer electrolytes for high energy density ternary cathode material-based lithium batteries. *Electrochem. Energy Rev.* **2019**, *2*, 128–148. [CrossRef]
16. Zhang, Q.; Liu, K.; Ding, F.; Liu, X. Recent advances in solid polymer electrolytes for lithium batteries. *Nano Res.* **2017**, *10*, 4139–4174. [CrossRef]
17. Zhao, Q.; Liu, X.; Stalin, S.; Khan, K.; Archer, L.A. Solid-state polymer electrolytes with in-built fast interfacial transport for secondary lithium batteries. *Nat. Energy* **2019**, *4*, 365–373. [CrossRef]
18. Zhao, Y.; Wang, L.; Zhou, Y.; Liang, Z.; Tavajohi, N.; Li, B.; Li, T. Solid polymer electrolytes with high conductivity and transference number of Li ions for Li-based rechargeable batteries. *Adv. Sci.* **2021**, *8*, 2003675. [CrossRef]
19. Jiang, Y.; Yan, X.; Ma, Z.; Mei, P.; Xiao, W.; You, Q.; Zhang, Y. Development of the PEO Based Solid Polymer Electrolytes for All-Solid State Lithium Ion Batteries. *Polymers* **2018**, *10*, 1237. [CrossRef] [PubMed]
20. Cheng, Z.; Liu, T.; Zhao, B.; Shen, F.; Jin, H.; Han, X. Recent advances in organic-inorganic composite solid electrolytes for all-solid-state lithium batteries. *Energy Storage Mater.* **2021**, *34*, 388–416. [CrossRef]
21. Wan, J.; Xie, J.; Mackanic, D.G.; Burke, W.; Bao, Z.; Cui, Y. Status, promises, and challenges of nanocomposite solid-state electrolytes for safe and high performance lithium batteries. *Mater. Today Nano* **2018**, *4*, 1–16. [CrossRef]
22. Zhao, Q.; Stalin, S.; Zhao, C.-Z.; Archer, L.A. Designing solid-state electrolytes for safe, energy-dense batteries. *Nat. Rev. Mater.* **2020**, *5*, 229–252. [CrossRef]
23. Liu, W.; Liu, N.; Sun, J.; Hsu, P.C.; Li, Y.Z.; Lee, H.W.; Cui, Y. Ionic conductivity enhancement of polymer electrolytes with ceramic nanowire fillers. *Nano Lett.* **2015**, *15*, 2740–2745. [CrossRef] [PubMed]
24. Bae, J.; Li, Y.; Zhang, J.; Zhou, X.; Zhao, F.; Shi, Y.; Goodenough, J.B.; Yu, G. A 3D nanostructured hydrogel-framework-derived high-performance composite polymer lithium-Ion electrolyte. *Angew. Chem. Int. Ed.* **2018**, *57*, 2096–2100. [CrossRef]
25. Gong, Y.; Fu, K.; Xu, S.; Dai, J.; Hamann, T.R.; Zhang, L.; Hitz, G.T.; Fu, Z.; Ma, Z.; McOwen, D.W.; et al. Lithium-ion conductive ceramic textile: A new architecture for flexible solid-state lithium metal batteries. *Mater. Today* **2018**, *21*, 594–601. [CrossRef]
26. Lin, D.; Yuen, P.Y.; Liu, Y.; Liu, W.; Liu, N.; Dauskardt, R.H.; Cui, Y. A silica-aerogel-reinforced composite polymer electrolyte with high ionic conductivity and high modulus. *Adv. Mater.* **2018**, *30*, e1802661. [CrossRef]
27. Wang, Z.; Shen, L.; Deng, S.; Cui, P.; Yao, X. 10 μm-thick high-strength solid polymer electrolytes with excellent interface compatibility for flexible all-solid-state lithium-metal batteries. *Adv. Mater.* **2021**, *33*, 2100353. [CrossRef]
28. Alarco, P.J.; Abu-Lebdeh, Y.; Abouimrane, A.; Armand, M. The plastic-crystalline phase of succinonitrile as a universal matrix for solid-state ionic conductors. *Nat. Mater.* **2004**, *3*, 476–481. [CrossRef] [PubMed]
29. Gao, H.; Xue, L.; Xin, S.; Park, K.; Goodenough, J.B. A plastic-crystal electrolyte interphase for all-solid-state sodium batteries. *Angew. Chem. Int. Ed.* **2017**, *56*, 5541–5545. [CrossRef] [PubMed]
30. Jiang, T.; He, P.; Wang, G.; Shen, Y.; Nan, C.-W.; Fan, L.-Z. Solvent-free synthesis of thin, flexible, nonflammable garnet-based composite solid electrolyte for all-solid-state lithium batteries. *Adv. Energy Mater.* **2020**, *10*, 1903376. [CrossRef]
31. Pal, P.; Ghosh, A. Robust succinonitrile plastic crystal-based Ionogel for all-solid-state Li-ion and dual-ion batteries. *ACS Appl. Energy Mater.* **2020**, *3*, 4295–4304. [CrossRef]
32. Wei, T.; Zhang, Z.-H.; Wang, Z.-M.; Zhang, Q.; Ye, Y.-S.; Lu, J.-H.; Rahman, Z.U.; Zhang, Z.-W. Ultrathin solid composite electrolyte based on $Li_{6.4}La_3Zr_{1.4}Ta_{0.6}O_{12}$/PVDF-HFP/LiTFSI/succinonitrile for high-performance solid-state lithium metal batteries. *ACS Appl. Energy Mater.* **2020**, *3*, 9428–9435. [CrossRef]
33. Wu, X.-L.; Xin, S.; Seo, H.-H.; Kim, J.; Guo, Y.-G.; Lee, J.-S. Enhanced Li^+ conductivity in PEO-LiBOB polymer electrolytes by using succinonitrile as a plasticizer. *Solid State Ion.* **2011**, *186*, 1–6. [CrossRef]
34. Cui, Y.; Wan, J.Y.; Ye, Y.S.; Liu, K.; Chou, L.Y. A fireproof, lightweight, polymer-polymer solid-state electrolyte for safe lithium batteries. *Nano Lett.* **2020**, *20*, 1686–1692. [CrossRef] [PubMed]
35. Li, M.; Zhang, Z.; Yin, Y.; Guo, W.; Bai, Y.; Zhang, F.; Zhao, B.; Shen, F.; Han, X. Novel polyimide separator prepared with two porogens for safe lithium-ion batteries. *ACS Appl. Mater. Interfaces* **2020**, *12*, 3610–3616. [CrossRef]

36. Shen, F.; Wang, K.; Yin, Y.; Shi, L.; Zeng, D.; Han, X. PAN/PI functional double-layer coating for dendrite-free lithium metal anodes. *J. Mater. Chem. A* **2020**, *8*, 6183–6189. [CrossRef]
37. Wan, J.; Xie, J.; Kong, X.; Liu, Z.; Liu, K.; Shi, F.; Pei, A.; Chen, H.; Chen, W.; Chen, J.; et al. Ultrathin, flexible, solid polymer composite electrolyte enabled with aligned nanoporous host for lithium batteries. *Nat. Nanotechnol.* **2019**, *14*, 705–711. [CrossRef]
38. Wang, Q.; Yuan, B.; Lu, Y.; Shen, F.; Zhao, B.; Han, X. Robust and high thermal-stable composite polymer electrolyte reinforced by PI nanofiber network. *Nanotechnology* **2021**. [CrossRef]
39. Ye, Y.; Chou, L.-Y.; Liu, Y.; Wang, H.; Lee, H.K.; Huang, W.; Wan, J.; Liu, K.; Zhou, G.; Yang, Y.; et al. Ultralight and fire-extinguishing current collectors for high-energy and high-safety lithium-ion batteries. *Nat. Energy* **2020**, *5*, 786–793. [CrossRef]

 *polymers*

 MDPI

Article

Effect of Process Control Parameters on the Filtration Performance of PAN–CTAB Nanofiber/Nanonet Web Combined with Meltblown Nonwoven

Hyo Kyoung Kang [1,2], Hyun Ju Oh [1], Jung Yeon Kim [1], Hak Yong Kim [2,3,*] and Yeong Og Choi [1,*]

[1] Advanced Textile R&D Department, Korea Institute of Industrial Technology, Ansan 15588, Korea; hkkang@kitech.re.kr (H.K.K.); hjoh33@kitech.re.kr (H.J.O.); cobalt98@kitech.re.kr (J.Y.K.)
[2] Department of Organic Materials and Fiber Engineering, Jeonbuk National University, Jeonju 54896, Korea
[3] Department of Nano Convergence Engineering, Jeonbuk National University, Jeonju 54896, Korea
* Correspondence: dragon4875@gmail.com (H.Y.K.); yochoi@kitech.re.kr (Y.O.C.); Tel.: +82-(0)63-270-2351 (H.Y.K.); +82-(0)31-8040-6084 (Y.O.C.)

Citation: Kang, H.K.; Oh, H.J.; Kim, J.Y.; Kim, H.Y.; Choi, Y.O. Effect of Process Control Parameters on the Filtration Performance of PAN–CTAB Nanofiber/Nanonet Web Combined with Meltblown Nonwoven. *Polymers* **2021**, *13*, 3591. https://doi.org/10.3390/polym13203591

Academic Editors: Ick-Soo Kim, Motahira Hashmi and Sana Ullah

Received: 14 September 2021
Accepted: 14 October 2021
Published: 19 October 2021

Publisher's Note: MDPI stays neutral with regard to jurisdictional claims in published maps and institutional affiliations.

Abstract: Nanofibers have potential applications as filters for particles with diameters <10 μm owing to their large specific surface area, macropores, and controllable geometry or diameter. The filtration efficiency can be increased by creating nanonets (<50 nm) whose diameter is smaller than that of nanofibers. This study investigates the effect of process conditions on the generation of nanonet structures from a polyacrylonitrile (PAN) solution containing cation surfactants; in addition, the filtration performance is analyzed. The applied electrospinning voltage and the electrostatic treatment of meltblown polypropylene (used as a substrate) are the most influential process parameters of nanonet formation. Electrospun polyacrylonitrile–cetylmethylammonium bromide (PAN–CTAB) showed a nanofiber/nanonet structure and improved thermal and mechanical properties compared with those of the electrospun PAN. The pore size distribution and filter efficiency of the PAN nanofiber web and PAN–CTAB nanofiber/nanonet web with meltblown were measured. The resulting PAN–CTAB nanofiber/nanonet air filter showed a high filtration efficiency of 99% and a low pressure drop of 7.7 mmH$_2$O at an air flow rate of 80 L/min. The process control methods for the nanonet structures studied herein provide a new approach for developing functional materials for air-filtration applications.

Keywords: nanonet; polyacrylonitrile; surfactant; electrospinning; meltblown; nanofiber/nanonet

1. Introduction

Air pollution threatens human health and the environment worldwide [1]. The manufacturing, transportation, and construction industries are some of the major contributors of air pollution as they release toxic waste and fine particles into the atmosphere [2,3]. High concentrations of particulate matter (PM) in the air are a primary cause of respiratory diseases [4–6]. PM, which consists of extremely small particles and liquid droplets [7], is classified into coarse (2.5–10 μm, PM10), fine (0.1–2.5 μm, PM2.5), and ultrafine (<0.1 μm, PM0.1) PM [8]. In particular, PM2.5 can easily penetrate the human lungs and bronchi after a short period of exposure [9,10]. To prevent the inhalation or ingestion of PM2.5, researchers recommend the use of a mask filter for personal protection during outdoor activities; further, a window screen and purification filter should be used for indoor air quality protection.

Fibrous filters are generally commonly used for air filtration because they have a network support structure, can trap PM, and are made of tortuous pore channels for air transport [11]. In general, fibrous filters can be classified into meltblown (MB), spunbond, electrospun, and needle-punch filters [12]. Among them, conventional filters such as nonwoven needle-punch, glass-fiber, and spunbond filters have large fiber diameters and are suitable for use as coarse filters [13]. By contrast, meltblown filters and electrospun filters have small diameters and pores; therefore, they are used for fine particle filtration. MB nonwoven fabrics have a large surface area per unit weight and remarkable barrier properties, with a pore size of 0.5–10 μm and an average diameter of 1–2 μm [14]. To increase the filtration efficiency, the fiber diameter must be decreased; however, there are limitations in the equipment design and process [15].

Electrospinning is a simple and versatile top–down spinning method capable of producing nano-to-submicron nonwoven filters from polymer and polymeric mixture solutions [16–18]. Electrospun nonwoven fabrics have a large surface area and porosity, making them significantly more effective in filtering out particles than their conventional counterparts [19,20]. In previous studies, we produced nano-sized nonwoven fabric filters by electrospinning various polymers, namely poly(vinyl alcohol) [21], poly(vinyl alcohol)/lignin [22], polystyrene [23], polyacrylonitrile (PAN) [24], polylmethylmethacrylate (PMMA) [24], polyethersulfone (PES) [25], polyurethane (PU) [26], and poly(lactic acid) [27]. The use of electrospinning for the fabrication of nano-sized nonwoven fabric filters improved the filtration performance over that of micrometer-sized nonwoven filters. Researchers have been conducting studies to further improve the efficiency of nanofiber filters [1,28]. To further reduce the diameter of electrospun nanofibers, electrospinning/net generation technology is being studied by controlling the process parameters. For instance, Shichao et al. [28] produced a PMIA nanofiber/net air filter with a filtration efficiency of 99.9% at a pressure drop of 92 Pa via electrospinning by mixing a surfactant with poly(m-phenylene isophthalamide). In addition, polyurethane (PU) was electrospun into a 20-nm-diameter nonwoven piece of fabric with a wide range of Steiner tree structures. An air filter with filtering efficiencies of 99% and 99.73% for PM1–0.5 and PM2.5–1, respectively, at a pressure drop of 28 Pa was fabricated [2]. The effect of collector type on net formation was also studied [29]. Hsiao et al. [30] added a surfactant to poly(vinylidene fluoride) to fabricate a nanofiber/nanonet and revealed that the filtration efficiency of the nonwoven fabric was 99.985% at 32 L/min and the pressure drop was 66.7 Pa. Furthermore, the effect of applied voltage on the formation of nanonets was studied. Several studies were conducted on air filters with nonwoven fabrics made of nanofibers/nanonets, and they achieved high efficiencies and low pressure drops.

PAN is a polar polymer that has a high dipole moment of 3.6 D; therefore, it can trap particulate pollutants in the air relatively easily [19,31]. In addition to a high filtration efficiency, it exhibits good ultraviolet resistance, weatherability, and chemical resistance [19]. PAN electrospun nanofiber membranes with these advantages are good materials for air filter media. As a related study, Lakshmanan et al. [32] produced an air filter media laminated with an electrospun web by adding a surfactant to the PAN solution. However, a nanonet structure was not observed. In fact, there are few studies related to the formation of nanonet structures in PAN polymers. Thus, there is substantial scope for manufacturing functional filter media by taking advantage of the characteristics of PAN and nanonets.

In this study, we investigated the formation conditions of nanonet structures in PAN nanofibers to obtain optimal filter media. The nanofiber–nanonet structure was achieved by controlling electrospinning process parameters such as the cationic additive, applied voltage, and electrostatically treated MB (e-MB) substrate (Figure 1). The morphology, thermal properties, and mechanical properties of the PAN nanofibers and PAN-CTAB nanofiber/nanonet web were analyzed. The pore size distribution and filter efficiency of PAN-CTAB/MB and pristine PAN/MB filter media were measured by varying the nanofiber basis weights. After fabricating a face mask using the PAN-CTAB nanofiber/nanonet web, the filtration efficiency and pressure drop were characterized under an increasing flow rate.

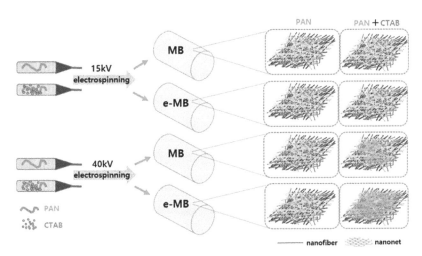

Figure 1. Schematic of the fabrication of PAN–CTAB nanofiber/nanonet web on electrostatically treated meltblown (e-MB) and discharged MB nonwoven filter.

2. Materials and Methods

2.1. Materials

Analytical-grade polyacrylonitrile (PAN, average Mw = 150,000 g mol^{-1}) was purchased from Sigma-Aldrich. Cetylmethylammonium bromide (CTAB), N,N-dimethyl formamide (DMF), and isopropyl alcohol (IPA) were obtained from Dae-Jung Chemical & Metals Co., Ltd., Siheung, Korea. The e-MB material used as a substrate was provided by Cheong-Jung clean Co., Ltd., Jincheon, Korea. All chemicals were used without further treatment.

2.2. Preparation of Filter Media

The PAN solution was prepared by dissolving 1.739 g of PAN in 20 mL of DMF and stirring for 10 h at 26 °C. The PAN–CTAB solution was prepared by adding 1.3 wt% of CTAB powder to the PAN solution and stirred for 12 h to form electrospinning solutions with increased conductivity [33]. Thus, two batches of solutions were prepared, PAN and PAN–CTAB. Subsequently, the solutions were poured into a 23-gauge syringe connected to a metal nozzle. The e-MB and discharged MB materials were used as substrates by sticking them on a drum collector. The e-MB material was immersed in IPA for 20 min, and then dried at room temperature for 24 h to nullify its electrostatic force. The electrospinning process conditions were as follows: the distance between the spinneret and the nonwoven substrates was 15 cm, the applied voltage from the power supply was 15 and 40 kV, the feed rate of the spinning solution was 5 μL/min, and the relative humidity (RH) and temperature were 40% ± 5%, and 22 ± 2 °C, respectively. Finally, the nanofiber web was dried in an oven at 100 °C for 2 h to evaporate the residual solvent.

2.3. Characterization of Nanofiber/Nanonet Web

The morphological characteristics of the PAN nanofiber web and PAN–CTAB nanofiber/nanonet web obtained with different voltages and substrates were investigated by field-emission scanning electron microscopy (FE-SEM, SU8010, Hitachi Co., Tokyo, Japan) at an acceleration voltage of 10 kV. The thermal properties of the two webs were analyzed by thermogravimetry (TGA, Q500, TA Instruments Co., New Castle, DE, USA) in an N$_2$ atmosphere with a heating rate of 10 °C/min. The thermal behaviors were examined by differential scanning calorimetry (DSC, 404 C, Netzsch Co., Selb, Germany) from 30 to 350 °C with a heating rate of 10 °C/min under an N$_2$ atmosphere. The crystal formation and structural characteristics due to the addition of CTAB were analyzed by X-ray diffraction (XRD, X'PERT-PRO Powder, PANalytical Co., Almelo, The Netherlands). The X-rays were generated at 40 kV and 30 mA with Cu–Ka radiation (wavelength, $\lambda = 0.154$ nm). The scan speed and measuring range were $4°$/min and $10°$–$60°$, respectively. The structural characteristics of the two types of webs were analyzed by Fourier transform infrared spectroscopy (FT-IR spectrometer, Nicolet NEXUS, Thermo Fisher Scientific Co., Waltham, MA, USA) using the ATR mode in the scan range of 400–4000 cm^{-1}. The mechanical properties with and without CTAB were determined using a universal testing machine (Instron3343, Instron Co., Barcelona, Spain) equipped with a 50 N load cell. At least 10 rectangular web specimens (50 mm $\times$ 100 mm) were stretched at a constant crosshead speed of 10 mm/min and tested.

2.4. Analysis of Filter Media

The pore-size distributions and mean pore diameters of the PAN and PAN–CTAB filter media samples were measured by capillary flow porometry (CFP, CFO-1500AEX, Porous Materials Inc., Ithaca, NY, USA). All the samples were cut into 3 cm $\times$ 3 cm squares. Galwick with a surface tension of 20.1 dynes/cm was used as the wetting agent. Maximum volumetric flow rates of 20,000 and 100,000 L/m were used in the air permeability and pore distribution tests. The filtration performance of the media with the MB nonwoven material was tested using an automated filter tester (TSI 8130, TSI Instruments Co. Ltd., Shoreview, MN, USA). NaCl aerosols with a mass median diameter of 0.26 mm were produced using an atomizing air pump. The aerosol particles permeated the sample at a flow rate of 32 L/min. The filter media had an effective test area of 100 cm^2. A mask was fabricated by adding spunbond nonwoven fabrics, which did not affect the filtering efficiency of the samples electrospun on the MB nonwoven fabrics. The size of mask sample was a width of 18 cm and a height of 17 cm, and the result was obtained by folding the middle part thrice. The filtering performance was tested with different air volumetric flow rates (10–80 L/min). Each sample was tested five times to ensure accuracy.

3. Results and Discussion

A nanofiber filter media with nanonet structures was manufactured to improve its ability to filter ultrafine particles. To manufacture the PAN nanofiber web with nanonet structures, the effects of an ionic surfactant and spinning conditions were confirmed. Figure 2 illustrates the FE-SEM images of the PAN nanofiber web surface prepared according to the surfactant, substrate conditions (charged and uncharged MB substrates), and spinning voltages (15 and 40 kV). Figure 2a–d shows the surface images of the pristine PAN nanofiber web according to the manufacturing conditions. The formation of a nanonet under the applied voltage and substrate conditions was not observed. Figure 2e–f depicts the surface images of the PAN–CTAB nanofiber web or nanofiber/nanonet web prepared by adding CTAB as the surfactant. A nanonet was observed at an applied voltage ≥ 40 kV. The PAN–CTAB nanofiber/nanonet web spun on the e-MB substrate (Figure 2h) had a larger nanonet than that of the web spun on the discharged MB substrate. This result led to the conclusion that the ions in the e-MB substrate influence the nanonet production under high-voltage conditions (40 kV). The pristine PAN nanofiber web had an average diameter of 245.3 nm.

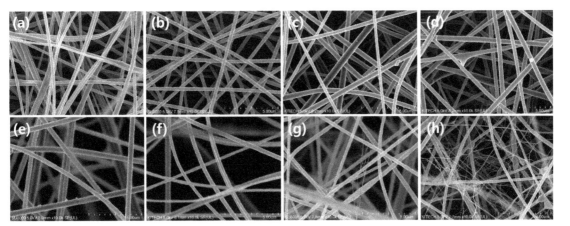

Figure 2. Field-emission scanning electron microscopy (FE-SEM) images of PAN (**a–d**) and PAN–CTAB (**e–h**) nanofiber/(nanonet) web with varying spinning voltage and substrate type: (**a,e**) 15 kV, meltblown (MB), (**b,f**) 15 kV, e-MB, (**c,g**) 40 kV, MB, and (**d,g**) 40 kV, e-MB.

The average diameter of the PAN–CTAB nanofiber/nanonet web (Figure 2h) was 276.1 nm/26.1 nm. The addition of CTAB increased the diameter of the PAN nanofibers compared to that in the PAN nanofiber web.

TGA and DSC were performed to evaluate the effect of the thermal properties of CTAB on the PAN–CTAB nanofiber/nanonet web. The results of the TGA are illustrated in Figure 3a. The two types of nanofibers and CTAB powder obtained under each of the conditions began to decompose at 300.9, 229.18, and 284.81 °C. It was confirmed that the addition of CTAB lowered the temperature of thermal decomposition of the PAN–CTAB nanofiber/nanonet web. The DSC analysis confirmed the effect of CTAB powder on the thermal properties of the PAN nanofiber web. In general, the PAN nanofiber web generates exothermic peaks owing to cyclization, dehydration, and oxidation reactions at 200–350 °C [33]. During the DSC analysis, the C≡N bonds were converted to C=N bonds with the stabilization process [34]. The exothermic peaks of the PAN nanofiber web and PAN–CTAB nanofiber/nanonet web were confirmed at 295 and 300.2 °C, respectively (Figure 3b). The endothermic peak of the CTAB powder appeared at approximately 102 °C, and the presence of CTAB was confirmed by the occurrence of an endothermic peak at the same temperature in the PAN–CTAB nanofiber/nanonet web. The onset temperature of the PAN nanofiber web increased by approximately 5 °C with the addition of CTAB. CTAB acted as an inhibitor to retard the cyclization reaction of PAN in the PAN–CTAB nanofiber/nanonet web, and blocked the recombination reaction by reducing the free-radical formation rate of the nitrile group [34].

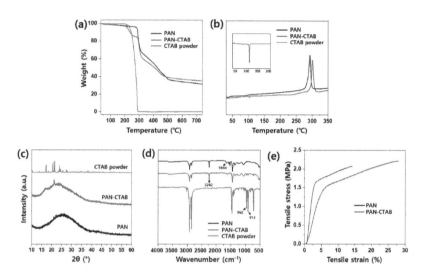

Figure 3. Characterizations of PAN nanofiber web, PAN–CTAB nanofiber/nanonet web, and CTAB powder. (**a**) Thermogravimetric analysis (TGA), (**b**) differential scanning calorimetry (DSC), (**c**) X-ray diffraction (XRD) patterns, and (**d**) Fourier transform infrared (FT-IR) spectra results of the three groups, respectively. (**e**) S–S curves of the PAN nanofiber and PAN–CTAB nanofiber/nanonet web.

Figure 3c shows the XRD patterns of the pristine PAN nanofiber web, PAN–CTAB nanofiber/nanonet web, and CTAB powder. The PAN nanofiber web displayed a broad crystal peak at 25.7°. The crystal peak of the CTAB powder exhibited high intensities at 17.3, 20.4, 21.4, and 24.3°. The PAN–CTAB nanofiber/nanonet web showed a broad peak at approximately 17–25° under the influence of CTAB. It is plausible that the addition of the CTAB powder enlarged the PAN nanofiber web crystals.

Figure 3d displays the FT-IR spectrum of the pristine PAN nanofiber web, CTAB powder, and PAN/CTAB nanofiber/nanonet web. The spectrum corresponding to the pristine CTAB exhibits the characteristic absorption bands of the CTAB functional groups, as reported previously [6]. The absorption bands between 2960 and 2820 and 1510 and 1440 cm^{-1} correspond to –CH$_2$– stretching and –CH$_2$– bending in CATB and the PAN–CTAB nanofiber/nanonet web, respectively, confirming the presence of long alkyl groups of CTAB in the PAN–CTAB nanofiber/nanonet web [35]. The pristine PAN nanofiber web exhibited prominent peaks at 1452, 1664, 2240, and 2923 cm^{-1}. The peaks at 1664, 2240, and 2923 cm^{-1} represent the stretching vibrations of –C=N, –C≡N, and –CH groups, respectively, whereas that at 1452 cm^{-1} represents the bending vibration of the –CH$_2$ group. A subtle increase in the characteristic peak of the PAN nanofiber web was observed at 1664 cm^{-1} with the addition of CTAB in the PAN nanofiber web.

Figure 3e summarizes the mechanical properties of the pristine PAN nanofiber web and PAN–CTAB nanofiber/nanonet web. The tensile strengths of the two webs were approximately 2.07 and 2.27 MPa, respectively. The tensile strength of the PAN–CTAB nanofiber/nanonet web improved by approximately 110% with the addition of the CTAB powder over that of the pristine web. The elongation of the PAN–CTAB nanofiber nanonet web also improved by approximately 200% compared with that of the pristine PAN nanofiber web. Notably, the tensile strain nearly doubled because the nanonet was distributed between nanofibers and served as a reinforcement in the cross-sectional direction [2].

Figure 4a,b shows the pore-size distribution of the filter media fabricated by laminating the PAN nanofiber web and PAN–CTAB nanofiber/nanonet web by weight (0.6–2.4 g/m^2) on the MB substrate. As the weight of the nanofiber web increases, leading to thickening of the nanofibers, the average pore size decreases as the pores are gradually blocked. In addition, in the case of the PAN–CTAB filter media with the net structure, the pore-size distribution was confirmed to be narrower than that of the PAN filter media owing to nanonet generation. The inset graph, an enlargement of the region corresponding to <10 μm diameter in the original graph, shows that the smallest pore size is attained at 0.9 g/m² in the case of the PAN–CTAB nanofiber/nanonet web.

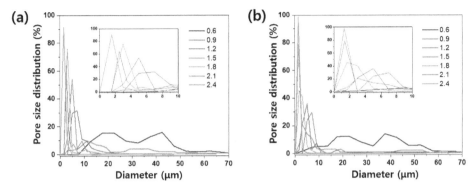

Figure 4. Pore-size distribution of (**a**) PAN and (**b**) PAN–CTAB filter media with varying base weight of the electrospun web.

Despite its high efficiency, the application of a nanofiber filter as a mask is limited by its high breathing resistance. The filtration efficiency was evaluated because it was affected by the formation of the nanonet structure in the nanofiber. Figure 5a shows the filtering efficiency according to the nanofiber weight. The control MB group without nanofibers had a filtering efficiency of 53.8%, and the filter media laminated with the PAN nanofiber web had an increased filtration efficiency of 72.4–88.7% as the weight of the nanofiber increased to 0.3–2.4 g/m². The efficiency of the filter media with stacked PAN–CTAB nanofiber/nanonet webs also improved to 74.9–99%. As the filter efficiency increased, the pressure drop also increased to 0.489 (only MB) up to 3.2 mmH₂O for the media laminated with the PAN nanofiber web, and to 0.489 (only MB) up to 3.7 mmH₂O for the media laminated with the PAN–CTAB nanofiber/nanonet web. Based on these results, the quality factor (QF = −ln(1 − η)/Δp, where η is the filtration efficiency, and Δp is the pressure drop) was calculated to evaluate the comprehensive filtering performance [36].

The QF is an indicator of the filtering efficiency considered for the pressure drop in the filter media. The filter in which the PAN–CTAB nanofiber/nanonet web was stacked from 0.9 g/m² or higher by weight of the laminated nanofiber had a significantly higher QF. The SEM images in Figure 5d,e confirm NaCl aerosol capture on the nanofiber after the filtration test. A notably larger amount of NaCl was adsorbed to the filter media in which the PAN–CTAB nanofiber/nanonet web was laminated than that on which the pristine PAN nanofiber web was laminated. This is because its ability to trap particles was improved by the presence of the nanonet between large pores.

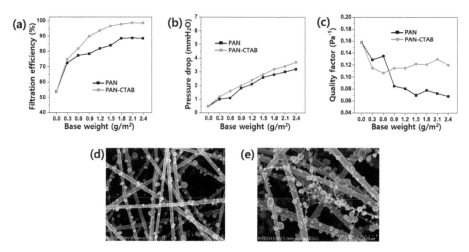

Figure 5. (**a**) Filtration efficiency, (**b**) pressure drop, and (**c**) quality factor of the PAN and PAN–CTAB filter media for different base weights of the electrospun web. The SEM images of (**d**) PAN nanofiber and (**e**) PAN–CTAB filter media after five iterations of the TSI8130 filtration evaluation test.

A mask prototype was fabricated with a PAN–CTAB nanofiber/nanonet web-layered sample of 2.1 g/m^2, which had the highest QF value (0.13 Pa^{-1}) among the filter media. As shown in the schematic diagram of Figure 6a, the mask was prepared by adding spunbond nonwoven fabric on both sides of the sample, which did not affect the filtering efficiency (Figure 6b). The filtering efficiency was tested with various flow rates. Consequently, the pressure drop across the layers of the mask gradually increased at high flow rates, and the efficiency was maintained at 98.5–99%. When the flow rate reached 80 L/min, the pressure drop was only 7.7 mmH_2O, which indicates high efficiency.

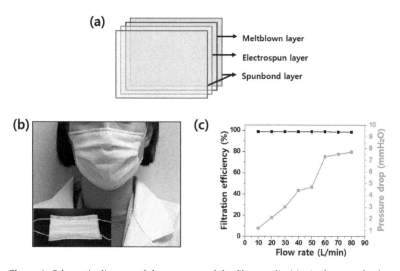

Figure 6. Schematic diagram of the structure of the filter media (**a**). A photograph of a person wearing the fabricated mask (**b**). Filtration performance evaluation by flow rate (**c**).

4. Conclusions

PAN–CTAB filter media containing a nanofiber/nanonet web were prepared using a cationic surfactant to increase the ion mobility via electrospinning at a high voltage. A larger nanonet was formed on an MB nonwoven substrate when the applied electrospinning voltage was 40 kV. The residual amount at 750 °C was found to be 4.5% more than that of the PAN–CTAB nanofiber/nanonet web. The DSC results showed that the addition of CTAB improved the thermal properties of the web as the exothermic peak temperatures increased by 5 °C. In the sample with the nanonet formed by the addition of CTAB, application of high voltage (40 kV), and use of the e-MB substrate, the nanofiber diameter increased; in addition, the tensile strength and tensile strain increased by 110% and 200%, respectively. The filtration efficiencies of the PAN (nanofiber web laminated on MB) and PAN–CTAB filter media were 88.7% and 99%, respectively, at the largest weight (nanofiber/nanonet web). The sample with the highest QF (0.13 Pa^{-1}) among the filter media was a 2.1 g/m^2 stack of PAN–CTAB nanofiber/nanonet web. A high efficiency, low-pressure-drop (3.4 mmH$_2$O) filter, with a filtration efficiency of 99%, was achieved. For application in a face mask, the uniformity of human inhalation and exhalation was considered, and a high filtration efficiency was measured at a high flow rate (80 L/min). The PAN–CTAB filter media have markedly improved properties compared to those of the PAN filter media of the control groups. In summary, the specifications of the PAN nanofiber/nanonet web can be altered by controlling the process parameters.

Author Contributions: Conceptualization, H.K.K., H.Y.K. and Y.O.C.; methodology and investigation, H.K.K. and H.J.O.; formal analysis, H.K.K., J.Y.K. and H.J.O.; writing—original draft preparation, H.K.K., J.Y.K. and H.J.O.; writing—review and editing, H.K.K., H.J.O., H.Y.K. and Y.O.C.; supervision, H.Y.K. and Y.O.C.; project administration, H.K.K., J.Y.K. and Y.O.C.; funding acquisition, Y.O.C. All authors have read and agreed to the published version of the manuscript.

Funding: This research was funded by the Korea Institute of Industrial Technology, grant number EH210009.

Institutional Review Board Statement: Not applicable.

Informed Consent Statement: Not applicable.

Data Availability Statement: Not applicable.

Conflicts of Interest: The authors declare no conflict of interest.

References

1. Zhu, M.; Han, J.; Wang, F.; Shao, W.; Xiong, R.; Zhang, Q.; Pan, H.; Yang, Y.; Samal, S.K.; Zhang, F.; et al. Electrospun nanofibers membranes for effective air filtration. *Macromol. Mater. Eng.* **2017**, *302*, 1–27. [CrossRef]
2. Zuo, F.; Zhang, S.; Liu, H.; Fong, H.; Yin, X.; Yu, J.; Ding, B. Free-standing polyurethane nanofiber/nets air filters for effective PM capture. *Small* **2017**, *13*, 1702139. [CrossRef]
3. Brook, R.D.; Rajagopalan, S.; Pope, C.A.; Brook, J.R.; Bhatnagar, A.; Diez-Roux, A.V.; Holguin, F.; Hong, Y.; Luepker, R.V.; Mittleman, M.A.; et al. Particulate matter air pollution and cardiovascular disease: An update to the scientific statement from the american heart association. *Circulation* **2010**, *121*, 2331–2378. [CrossRef]
4. Heft-Neal, S.; Burney, J.; Bendavid, E.; Burke, M. Robust relationship between air quality and infant mortality in Africa. *Nature* **2018**, *559*, 254–258. [CrossRef]
5. Zhang, Q.; Jiang, X.; Tong, D.; Davis, S.J.; Zhao, H.; Geng, G.; Feng, T.; Zheng, B.; Lu, Z.; Streets, D.G.; et al. Transboundary health impacts of transported global air pollution and international trade. *Nature* **2017**, *543*, 705–709. [CrossRef]
6. Dockery, D.W.; Pope, C.A.; Xu, X.; Spengler, J.D.; Ware, J.H.; Fay, M.E.; Benjamin, G.; Ferris, J.; Speizer, F.E. An association between air pollution and mortality in six U.S. cities. *N. Engl. J. Med.* **1993**, *29*, 1753–1759. [CrossRef]
7. Wang, X.; Xiang, H.; Song, C.; Zhu, D.; Sui, J.; Liu, Q.; Long, Y. Highly efficient transparent air filter prepared by collecting-electrode-free bipolar electrospinning apparatus. *J. Hazard. Mater.* **2020**, *385*, 121535. [CrossRef] [PubMed]
8. Senlin, L.; Zhenkun, Y.; Xiaohui, C.; Minghong, W.; Guoying, S.; Jiamo, F.; Paul, D. The relationship between physicochemical characterization and the potential toxicity of fine particulates (PM$_{2.5}$) in Shanghai atmosphere. *Atmos. Environ.* **2008**, *42*, 7205–7214. [CrossRef]
9. Shi, Y.; Ji, Y.; Sun, H.; Hui, F.; Hu, J.; Wu, Y.; Fang, J.; Lin, H.; Wang, J.; Duan, H.; et al. Nanoscale characterization of PM 2.5 airborne pollutants reveals high adhesiveness and aggregation capability of soot particles. *Sci. Rep.* **2015**, *5*, 1–10. [CrossRef]

10. Zhang, L.; Jin, X.; Johnson, A.C.; Giesy, J.P. Hazard posed by metals and As in PM2.5 in air of five megacities in the Beijing-Tianjin-Hebei region of China during APEC. *Environ. Sci. Pollut. Res.* **2016**, *23*, 17603–17612. [CrossRef] [PubMed]
11. Barhate, R.S.; Ramakrishna, S. Nanofibrous filtering media: Filtration problems and solutions from tiny materials. *J. Memb. Sci.* **2007**, *296*, 1–8. [CrossRef]
12. Zhang, H.; Liu, J.; Zhang, X.; Huang, C.; Jin, X. Design of electret polypropylene melt blown air filtration material containing nucleating agent for effective PM2.5 capture. *RSC Adv.* **2018**, *8*, 7932–7941. [CrossRef]
13. Zhang, H.; Liu, J.; Zhang, X.; Huang, C.; Zhang, Y.; Fu, Y.; Jin, X. Design of three-dimensional gradient nonwoven composites with robust dust holding capacity for air filtration. *J. Appl. Polym. Sci.* **2019**, *136*, 47827. [CrossRef]
14. Kim, H.J.; Han, S.W.; Joshi, M.K.; Kim, C.S. Fabrication and characterization of silver nanoparticle-incorporated bilayer electrospun–melt-blown micro/nanofibrous membrane. *Int. J. Polym. Mater. Polym. Biomater.* **2017**, *66*, 514–520. [CrossRef]
15. Dutton, K.C. Overview and analysis of the meltblown process and parameters. *J. Text. Apparel. Technol. Manag.* **2008**, *6*, 1–24.
16. Joshi, M.K.; Tiwari, A.P.; Pant, H.R.; Shrestha, B.K.; Kim, H.J.; Park, C.H.; Kim, C.S. In situ generation of cellulose nanocrystals in polycaprolactone nanofibers: Effects on crystallinity, mechanical strength, biocompatibility, and biomimetic mineralization. *ACS Appl. Mater. Interfaces* **2015**, *7*, 19672–19683. [CrossRef]
17. Fibers, N.; Bognitzki, B.M.; Czado, W.; Frese, T.; Schaper, A.; Hellwig, M.; Steinhart, M.; Greiner, A.; Wendorff, J.H. Nanostructured fibers via electrospinning **. *Adv. Mater.* **2001**, *13*, 70–72. [CrossRef]
18. Long, Y.; Li, M.; Gu, C.; Wan, M.; Duvail, J.; Liu, Z.; Fan, Z. Progress in Polymer Science Recent advances in synthesis, physical properties and applications of conducting polymer nanotubes and nanofibers. *Prog. Polym. Sci.* **2011**, *36*, 1415–1442. [CrossRef]
19. Liu, C.; Hsu, P.; Lee, H.; Ye, M.; Zheng, G.; Liu, N.; Li, W.; Cui, Y. Transparent air filter for high-efficiency PM$_{2.5}$ capture. *Nat. Commun.* **2015**, *6*, 6205. [CrossRef]
20. Kaur, S.; Sundarrajan, S.; Rana, D.; Matsuura, T.; Ramakrishna, S. Influence of electrospun fiber size on the separation efficiency of thin film nanofiltration composite membrane. *J. Memb. Sci.* **2012**, *392–393*, 101–111. [CrossRef]
21. Qin, X.; Wang, S. Electrospun nanofibers from crosslinked poly(vinyl alcohol) and its filtration efficiency. *J. Appl. Polym. Sci.* **2008**, *109*, 951–956. [CrossRef]
22. Cui, J.; Lu, T.; Li, F.; Wang, Y.; Lei, J.; Ma, W.; Zou, Y.; Huang, C. Flexible and transparent composite nanofiber membrane that was fabricated via a "green" electrospinning method for efficient particulate matter 2.5 capture. *J. Colloid Interface Sci.* **2021**, *582*, 506–514. [CrossRef] [PubMed]
23. Lin, T.; Wang, H.; Wang, H.; Wang, X. The charge effect of cationic surfactants on the elimination of fibre beads in the electrospinning of polystyrene. *Nanotechnology* **2004**, *15*, 1375–1381. [CrossRef]
24. Myoung, K.; Bagus, A.; Iskandar, F.; Bao, L.; Niinuma, H.; Okuyama, K. Morphology optimization of polymer nanofiber for applications in aerosol particle filtration. *Sep. Purif. Technol.* **2010**, *75*, 340–345. [CrossRef]
25. Sambaer, W.; Zatloukal, M.; Kimmer, D. 3D modeling of filtration process via polyurethane nanofiber based nonwoven filters prepared by electrospinning process. *Chem. Eng. Sci.* **2011**, *66*, 613–623. [CrossRef]
26. Wang, Z.; Pan, Z. Preparation of hierarchical structured nano-sized/porous poly(lactic acid) composite fibrous membranes for air filtration. *Appl. Surf. Sci.* **2015**, *356*, 1168–1179. [CrossRef]
27. Greiner, A.; Wendorff, J.H. Electrospinning: A fascinating method for the preparation of ultrathin fibers. *Angew. Chem.-Int. Ed.* **2007**, *46*, 5670–5703. [CrossRef]
28. Zhang, S.; Liu, H.; Yin, X.; Li, Z.; Yu, J.; Ding, B. Tailoring mechanically robust poly(m-phenylene isophthalamide) nanofiber/nets for ultrathin high-efficiency air filter. *Sci. Rep.* **2017**, *7*, 40550. [CrossRef]
29. Zhang, S.; Liu, H.; Tang, N.; Ge, J.; Yu, J.; Ding, B. Direct electronetting of high-performance membranes based on self-assembled 2D nanoarchitectured networks. *Nat. Commun.* **2019**, *10*, 1458. [CrossRef]
30. Li, X.; Wang, C.; Huang, X.; Zhang, T.; Wang, X.; Min, M.; Wang, L.; Huang, H.; Hsiao, B.S. Anionic surfactant-triggered steiner geometrical poly(vinylidene fluoride) nanofiber/nanonet air filter for efficient particulate matter removal. *ACS Appl. Mater. Interfaces* **2018**, *10*, 42891–42904. [CrossRef]
31. Su, J.; Yang, G.; Cheng, C.; Huang, C.; Xu, H.; Ke, Q. Hierarchically structured TiO$_2$/PAN nanofibrous membranes for high-efficiency air filtration and toluene degradation. *J. Colloid Interface Sci.* **2017**, *507*, 386–396. [CrossRef]
32. Lakshmanan, A.; Gavali, D.S.; Thapa, R.; Sarkar, D. Synthesis of CTAB-Functionalized Large-Scale Nanofibers Air Filter Media for Efficient PM2.5 Capture Capacity with Low Airflow Resistance. *ACS Appl. Polym. Mater.* **2021**, *3*, 937–948. [CrossRef]
33. Lu, X.; Wang, C.; Wei, Y. One-dimensional composite nanomaterials: Synthesis by electrospinning and their applications. *Small* **2009**, *5*, 2349–2370. [CrossRef] [PubMed]
34. Ji, L.; Saquing, C.; Khan, S.A.; Zhang, X. Preparation and characterization of silica nanoparticulate-polyacrylonitrile composite and porous nanofibers. *Nanotechnology* **2008**, *19*, 085605. [CrossRef] [PubMed]
35. Liu, G.Q.; Jin, Z.G.; Liu, X.X.; Wang, T.; Liu, Z.F. Anatase TiO2 porous thin films prepared by sol-gel method using CTAB surfactant. *J. Sol.-Gel. Sci. Technol.* **2007**, *41*, 49–55. [CrossRef]
36. Hung, C.H.; Leung, W.W.F. Filtration of nano-aerosol using nanofiber filter under low Peclet number and transitional flow regime. *Sep. Purif. Technol.* **2011**, *79*, 34–42. [CrossRef]

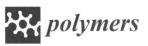

Article

Fabrication and Characterization of Electrospun Folic Acid/Hybrid Fibers: In Vitro Controlled Release Study and Cytocompatibility Assays

Fatma Nur Parın [1,*], Sana Ullah [2], Kenan Yıldırım [1], Motahira Hashmi [2] and Ick-Soo Kim [2,*]

1   Faculty of Engineering and Nature Science, Department of Polymer Materials Engineering,
    Mimar Sinan Campus, Bursa Technical University, Bursa 16310, Turkey; kenan.yildirim@btu.edu.tr
2   Nano Fusion Technology Research Group, Division of Frontier Fibers, Institute for Fiber Engineering (IFES),
    Interdisciplinary Cluster for Cutting Edge Research (ICCER), Shinshu University, Ueda 386-8567, Japan;
    sanamalik269@gmail.com (S.U.); motahirashah31@gmail.com (M.H.)
*   Correspondence: nur.parin@btu.edu.tr (F.N.P.); Kim@shinshu-u.ac.jp (I.-S.K.)

Citation: Parın, F.N.; Ullah, S.; Yıldırım, K.; Hashmi, M.; Kim, I.-S. Fabrication and Characterization of Electrospun Folic Acid/Hybrid Fibers: In Vitro Controlled Release Study and Cytocompatibility Assays. Polymers 2021, 13, 3594. https://doi.org/10.3390/polym13203594

Academic Editor: Deng-Guang Yu

Received: 20 September 2021
Accepted: 13 October 2021
Published: 19 October 2021

Abstract: The fabrication of skin-care products with therapeutic properties has been significant for human health trends. In this study, we developed efficient hydrophilic composite nanofibers (NFs) loaded with the folic acid (FA) by electrospinning and electrospraying processes for tissue engineering or wound healing cosmetic applications. The morphological, chemical and thermal characteristics, in vitro release properties, and cytocompatibility of the resulting composite fibers with the same amount of folic acid were analyzed. The SEM micrographs indicate that the obtained nanofibers were in the nanometer range, with an average fiber diameter of 75–270 nm and a good porosity ratio (34–55%). The TGA curves show that FA inhibits the degradation of the polymer and acts as an antioxidant at high temperatures. More physical interaction between FA and matrices has been shown to occur in the electrospray process than in the electrospinning process. A UV-Vis in vitro study of FA-loaded electrospun fibers for 8 h in artificial acidic (pH 5.44) and alkaline (pH 8.04) sweat solutions exhibited a rapid release of FA-loaded electrospun fibers, showing the effect of polymer matrix–FA interactions and fabrication processes on their release from the nanofibers. PVA-CHi/FA webs have the highest release value, with 95.2% in alkaline media. In acidic media, the highest release (92%) occurred on the PVA-Gel–CHi/sFA sample, and this followed first-order and Korsmeyer–Peppas kinetic models. Further, the L929 cytocompatibility assay results pointed out that all NFs (with/without FA) generated had no cell toxicity; on the contrary, the FA in the fibers facilitates cell growth. Therefore, the nanofibers are a potential candidate material in skin-care and tissue engineering applications.

Keywords: hybrid nanofiber; cytotoxicity; folic acid; in vitro study; drug release

1. Introduction

Nanofibers have played a key role in different areas of biomedical research, ranging from drug release to wound healing, and cell regeneration owing to their high specific surface area and capacity to adjust their properties by altering composition and production conditions [1–4]. The use of hydrophilic polymers in the fabrication of electrospun nanofibers has been proven to be beneficial in the development of fast-dissolving delivery systems with decreased drug–drug interactions [5–7]. Generally, a blended combination of natural and synthetic polymers is preferred. Synthetic polymers can offer proper mechanical strength while also enhancing electrospinnability. Natural polymers can promote cellular adhesion and thus better imitate the ECM environment [8]. Furthermore, the increased hydrophilicity supplied by natural polymers is favorable when carrying hydrophilic drugs. Nanofibrous scaffolds made of natural and synthetic polymers were shown to be suitable for designing drug release systems via the electrospinning of blended polymers or combination with bioactive agents [9–14]. Previously, electrospun nanofibers have been produced from single polymer sources, and more recently, polymer combinations have been created to develop so-called polyblend nanofibers. Polyblend nanofibers are a new type of biomaterial that can function as mimics for native tissue while presenting topographical and biochemical stimuli that support healing [15–17].

Scaffolds are made from a variety of polymers, including poly(vinyl alcohol), poly(ethyl oxide), poly(lactic acid), poly(glycolic acid), poly(lactic-co-glycolic acid), poly(caprolactone), and natural extracellular matrix (ECM) analogs, such as collagen, gelatin, silk, chitin, chitosan, alginate, and hyaluronic acid, because of their lack of toxicity, superior biocompatibility, cell adhesion, and proliferation capability [18]. In this context, hydrophilic PVA-based materials can be blended with natural polymers and utilized in drug delivery systems [18–24].

Bioactive compounds can partially lose their therapeutic efficiency during the blend–electrospinning process. It is critical to have protective agents. To ensure that drugs stay bioactive, certain procedures must be performed [8]. In this context, electrospraying can be utilized for drug release applications, especially the encapsulation of bioactive compounds [25]. Electrospraying, also referred to as electro hydrodynamic atomization (EHDA), is usually presented as a unique and simple technology capable of controlling droplet formation electrostatically [26]. Folic acid (FA) is defined as a multifunctional, active substance that is safe to use and even combine with other chemicals [27]. Moreover, folic acid has received much attention from researchers in the fields of biomedical, bioengineering, and regenerative therapies, because of its non-immunogenicity, high stability, and ability to support tissue regeneration/repair [26,27]. In this regard, an increasing number of studies on the use of folic acid in micro/nanocapsule, liposome, hydrogel and/or nanofiber forms are available in a wide range of applications [28–30]. However, there are no studies in the literature on the utilization of nanofibers containing FA in tissue engineering applications, formed by both blending electrospinning and simultaneous (electrospinning and electrospraying) processes.

The main objective of this study is to fabricate various PVA-based nanofibers, including FA, via two distinct processes, electrospinning and simultaneous spinning (electrospinning and e-spraying), and then to evaluate their controlled release and cell growth characteristics. The properties of PVA-based NFs in terms of morphology, thermal resistance and chemical structure were analyzed by changing the type of polymer matrix. In addition, we assessed which had better cell growth and release values.

2. Materials and Methods

2.1. Materials

The Poly(vinyl alcohol) (PVA) (density 0.4–0.6 g/cm^3, purity 87.8% and M_w~30.000 g/mol) was purchased from ZAG Industrial Chemicals (Istanbul, Turkey). Gelatin from bovine skin (225 Bloom, type B, Bio Reagent) was supplied from Sigma Aldrich Chemical Company (St. Louis, MO, USA). Sodium alginate (medium molecular weight) and sodium bicarbonate (NaHCO$_3$) (99.7% purity) were purchased from Sigma Aldrich Chemical Company (St. Louis, MO, USA). Acetic acid (CH$_3$COOH) with 80% purity was also obtained from Sigma Aldrich Chemical Company (St. Louis, MO, USA), and ethanol (99.9% purity) was purchased from Tekkim Chemical Company (Bursa, Turkey). Chitosan (M_n 100,000–300,000) was supplied by Alfa Aesar Fine Chemicals and Metals (USA). Folic acid (C$_{19}$H$_{19}$N$_7$O$_6$) for biochemistry (98–102% purity) was provided by ChemSolute Company (Germany). The ethanol (99% purity) was purchased from Sigma Aldrich Chemical Company (St. Louis, MO, USA). L-Histidine mono-hydrochloride monohydrate (C$_6$H$_9$O$_2$N$_3$.HCI.H$_2$O) (>99% purity), sodium chloride (>99.5% purity), and sodium dihydrogen phosphate dihydrate (NaH$_2$PO$_4$.2H$_2$O) were supplied by Sigma Aldrich Chemical Company (St. Louis, MO, USA) to produce the artificial sweat solutions.

2.2. Methods

2.2.1. Fabrication of Neat Nanofibers and FA-Loaded Hybrid Nanofiber via Blending Electrospinning

Polymer solutions were prepared as in the previous study [10]. PVA granules were dissolved in distilled water at 90 °C until a 10% (w/w) homogenous PVA solution was produced. By stirring for 2 h at room temperature, gelatin was dissolved in distilled water and acetic acid binary-solvent systems (7:3 w/w) solution to get a 20% (w/w) gelatin solution. Alginate powders were dissolved in water to make a 2% (w/w) alginate solution. Chitosan powders were dissolved in acetic acid and distilled water binary-solvent systems (9:1 w/w) solution to obtain 7% chitosan solution. The process parameters and features of the composite polymer solutions are shown in Table 1 below. The average temperature was 25 °C, the % humidity was lower than 60, and the speed of the rotating collector was 150 rpm. In all FA-loaded polymer solutions, 22 mg of FA in 1 M 1 mL of $NaHCO_3$ was homogeneously added to the blend solutions.

Table 1. Process parameters for electrospinning.

Fiber Type	Polymer Ratio (*v/v*)	Voltage (kV)	Distance (mm)
PVA-Gel	4/1	27	75
PVA-CHi	4/1	24	97
PVA-Alg	4/1	26	90
PVA-Gel–CHi	3/1/1	26	78
PVA-Alg–CHi	3/1/1	26	78

The electrospinning of PVA-based composite nanofibers and FA-loaded PVA-based composite nanofibers was carried out by an electrospinning process (INOVENSO Nanospinner24 Device, Turkey). Neat polymer solutions without FA were prepared as a control. In the electrospinning process, the polymer solutions were transferred into plastic syringes (10 mL) and these syringes were attached to stainless nozzles with 0.5 mm inner diameters (Figure 1). Polymer solutions were dispensed into the rotary collector. Finally, neat and FA-loaded nanofibers were deposited in the collector due to the high electrical voltage.

2.2.2. Fabrication of FA-Spraying on Hybrid Nanofibers via Simultaneous Process

The simultaneous process (electrospinning + e-spraying process) has been detailed in a previous study [10]. Two syringe pumps for polymer solution and FA dispersion were utilized in the experiments. The environmental factors and the speed of the collector were the same as in blending electrospinning process. Schematic illustrations of the two distinct processes are shown in Figure 1. In the simultaneous process, all polymer solutions were prepared. Afterwards, the sonicated FA–ethanol solution and the polymer solutions were transferred into two distinct plastic syringes. These solutions were fed into the system by two distinct syringe pumps as in the electrospinning process.

The samples are encoded as polymer blend/FA and polymer blend/sFA for the electrospinning process and simultaneous process (electrospinning + e-spraying process), respectively.

2.2.3. Microstructural Analysis of the Hybrid Nanofibers

A Carl Zeiss/Gemini 300 Scanning Electron Microscope (SEM) was used to examine the morphology of all nanofibers (ZEISS Ltd., Jena, Germany). Before analysis, all samples were coated with gold for 20 min. The diameters of 60 individual fibers were randomly selected for each sample and measured using ImageJ (version 1.520 software).

2.2.4. Physical Analysis of Nanofibers

The ATR-FTIR spectra of nanofibers were acquired using a ThermoNicolet iS50 FT-IR (USA) spectrometer and an ATR (Attenuated Total Reflectance) adaptor (Smart Orbit

Diamond, MI, USA). Spectra in the range of 500–4000 cm^{-1} with automatic signal gain were collected (16 scans) at a 4 cm^{-1} resolution.

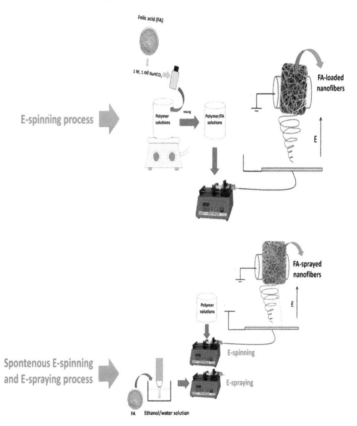

Figure 1. Schematic illustration of the processes.

2.2.5. Thermal Analysis of Nanofibers

The thermal analysis was performed using a TA/SDT650 TGA (USA). TGA analysis was carried out in a nitrogen (N$_{2(g)}$) atmosphere with a 20 °C min^{-1} heating rate and a temperature range of 30–600 °C, and in an oxygen (O$_{2(g)}$) atmosphere with a 20 °C min^{-1} heating rate and a temperature range of 600–900 °C.

2.2.6. Drug Release of Nanofibers and Kinetics

The ISO 105-E04:2013 procedure [31] was used to make the artificial sweat solutions. The total immersion procedure was performed to evaluate the vitamin-release behavior of FA-sprayed final fibers in acidic and alkaline sweat solutions at pH 5.44 and pH 8.04 [31]. Nanofibers of 30 cm^2 were placed in sealed glass tubes with 100 mL of acidic and alkaline sweat solution, respectively. After this, they were placed in a shaking incubator at 36 °C with 200 rpm shaking, and we applied the folic acid release profile. Samples of 3.5 mL were taken of the sweat solutions at the specified time intervals, and the associated absorbance value was determined in a UV-Vis spectrophotometer (Scinco/NEOYSY 2000, Seoul, South Korea) at max = 282 nm, which is folic acid's typical peak. The drug concentrations were determined using the calibration curve of a model vitamin prepared with a known concentration of folic acid in acidic and alkaline solutions. For the acidic solution, the calibration curve was calculated to be Y = 0.0486X + (−0.0402) (R^2 = 0.99992), and for alkaline solution, $Y = 0.0551X + (-0.0422)$ ($R^2 = 0.99959$), where X is the FA concentration (mg/L) and Y is the solution absorbance at 282 nm (linear range of 0.5–25 mg/L). UV-Vis spectroscopy was used to determine the amount of drug released. For each experiment, three replicates were obtained.

The kinetics of drug release were substantially controlled by polymer blends. The main mechanisms for drug release from a polymer matrix include drug diffusion and polymer relaxation/dissolution [32]. The cumulative release of FA (%) from all nanofibers was plotted with time. In this study, the kinetic mechanisms of all obtained nanofibers were calculated with zero-order, first-order, Hixson Crowell, Higuchi, and Korsmeyer–Peppas mathematical models. The different kinetic models are shown in Table 2.

Table 2. Kinetic models used to calculate FA release from the nanofibers.

Kinetic Model	Equation
Zero-order	$Q_t = k_0 t + Q_0$ Q_t: Drug concentration at time t Q_0: Initial drug concentration k_0: Zero order rate constant
First-order	$\log Q_t = \log Q_0 - kt /2303$ Q: Drug concentration released at time t Q_0: Initial drug concentration k: First order rate constant
Higuchi (Power Law)	$Q_t = k_h \, t^{1/2}$ Q: Drug concentration at time t k_h: Higuchi release rate constant
Hixson-Crowell	$Q_0^{1/3} - Q_t^{\frac{1}{3}} = k_h$ Q_0: Initial drug amount Qt: Drug amount at time t k: Hixson–Crowell rate constant
Korsmeyer–Peppas	$\frac{Qt}{Q\infty} = kt^n$ Qt: Drug concentration at time t $Q\infty$: Equilibrium drug concentration in the release medium $Qt/Q\infty$: Drug fraction in the release medium at time t k: Release rate constant n: Diffusional exponent

2.2.7. Entrapment Efficiency (EE) and FA Loading Capacity (LC) in Nanofibers

The FA entrapment efficiency (EE, %) was calculated as follows.

A previously known area of the fibrous materials (3×3 cm^2) was completely dissolved separately in acidic and alkaline sweat solutions. Afterward, about 3 mL of these solutions was evaluated at the FA characteristic wavelength using UV-Vis spectroscopy. The FA loading capacity of the nanofibers (LC, %) was measured to specify how much FA per polymer was entrapped in them. These equations are shown as Equations (1) and (2).

$$(100 - EE)\% = \frac{Amount \; of \; total \; FA \; release - Unentrapped \; FA \; amount}{Amount \; of \; total \; FA} \times 100 \quad (1)$$

$$LC = \frac{Amount \; of \; total \; FA \; release}{Initial \; amount \; of \; FA \; containing \; nanofiber} \times 100 \quad (2)$$

2.2.8. Cytotoxicity of Nanofibers
Preparation of Cell Culture

In the cell culture laboratory of the Biotechnology Department of the Bursa Technical University, all experimental aspects of the cell culture and cytotoxicity assessments were carried out in accordance with acceptable cell culture practices (ISO Standard) [32]. The L929 cell lines (mouse fibroblasts), as a reference cell line of the UNI EN ISO 10993-5 rule:

2009, were used to test medical products and materials for cytotoxicity [33]. L929 cells were seeded in T75 bottles with the addition of 10% FBS and 1% penicillin streptomycin to RPMI 1640. The cells were cultured in a humidified cell culture incubator at 37 °C in 5% CO_2 and observed daily with an inverted microscope and phase contrast (Olympus CKX41). When 80% confluence was detected, subculturing was carried out. In 96 tissue culture wells, a total of 5 cells/well were placed following disaggregation with trypsin/EDTA and the resuspension of cells in the media.

Preparation of Nanofiber Extract Solutions and MTT Assays

The ISO 10993-12: 2009 standard [34] was utilized in this study for the sample preparation of both the reference and the new nanofiber material. Both sides of standard-sized samples (3 cm^2 /mL) were sterilized in a laminar flow cabinet for 1 h under a UV light. In order to obtain extract solutions, 30 cm^2 (6 cm × 5 cm) sterile samples were placed in a 10 mL culture medium (RPMI 1640 with 1% Penicillin–Streptomycin, 10% serum) and cultured at 37 °C for 1 h. The samples were removed at the end of the extraction period, and the extract solutions were held at 37 °C during the cytotoxicity testing. MTT analysis was performed to determine the cytotoxicity of nanofiber extract solutions in the L929 cell line. In this regard, cells were exposed to the nanofiber extract solutions after 24 h of incubation of the cells seeded in 96-well tissue culture plates. Then, the media were aspirated to add MTT (5 mg/mL of stock of PBS) and the cells were incubated with MTT dye for a further 4 h (10 mL/well in 100 mL of cell suspension). Finally, the dye was carefully removed and 100 mL DMSO was added to each well. The absorption of the solution in each well was evaluated within a microplate reader at 570 nm. To assess cytotoxicity, the effects of 100% concentrations of original extract solutions from all reference and novel nanofiber samples on treated cells were compared to a negative control group that did not receive any chemicals. The average absorbance values and standard deviation values of living cells were calculated by averaging all the data obtained. Furthermore, cell viability in the control group was assumed to be 100%, and live cell percentages were assessed for all sample groups in comparison to the control.

2.3. Statistical Analysis

Statistical analysis of % cell viability values obtained after the cytocompability experiments were performed using IBM SPSS Statistics version 22 software (IBM software, Armonk, NY, USA). The results were statistically analyzed by Student–Newman–Keuls (SNK) test to determine the importance of the difference between different groups. The significance value level was identified as $p < 0.05$.

3. Results and Discussion

3.1. Morphology of the Hybrid Nanofibers

The neat blend fibers, FA-loaded composite fibers, and FA-sprayed composite fibers at the same quantities were successfully electrospun. The morphology of the fibers was characterized by SEM, as shown in Figure 2. To further analyze the diameter distributions of nanofibers, ImageJ software was used to determine the average diameter of 60 different and random nanofibers using SEM micrographs. Generally, the incorporation of FA decreases the average diameter of FA-loaded nanofibers. This could be due to the addition of FA into the polymer blend solutions, together with $NaHCO_3$, as a result of which the viscosity decreased.

It was noticed that the PVA-Gel, PVA-Alg, and PVA-CHi nanofibers were bead-free and smooth. Unlike PVA-Alg, which included more fiber interactivity and coarser fibers, PVA/Gel and PVA-CHi did not contain interacting or coarser fibers. An interconnected fiber structure was revealed in both neat PVA-Alg and PVA-Alg/FA fibers. However, in the PVA-Alg/FA sample, the fiber thickness was even thinner, and the porosity increased slightly (208.2 nm, 48.97%) compared to the neat one due to the FA-$NaHCO_{3(aq)}$ in neat PVA-Alg/FA fibers. As FA was sprayed on the fiber surfaces, morphology deteriorated

(Figure 2i), and film-like structures formed on the fibers' surfaces before the fibers dried. It is thought that this type of film formation is a consequence of the dissolution of the fiber with the solvent of the FA during the simultaneous process [10]. This heterogeneous appearance could also be an effect of controlled drug release. PVA-Gel fibers were dispersed and showed a tight structure with fewer fiber joints. On the other hand, FA-loading gave rise to larger fiber diameters and caused the fiber cross-section to adopt a flat form. After FA-loading, a partial film surface also formed, resulting from the viscosity decreasing due to the addition of the NaHCO$_3$ (Figure 2b). As seen in Figure 2c, the simultaneous process facilitated the formation of FA beads on the PVA-Gel fibers' surfaces, and/or layers, but the web morphology was destroyed as in the PVA-Alg/sFA and PVA-Gel/FA samples (Figure 2b,f). The SEM micrographs of the PVA-CHi nanofibers in a random spaghetti structure show that knotty fiber structures formed in some regions through a spontaneous process. Moreover, the fiber thickness increased from 96.26 nm to 101.86 nm, which occurred in parallel to porosity decline (Table 3).

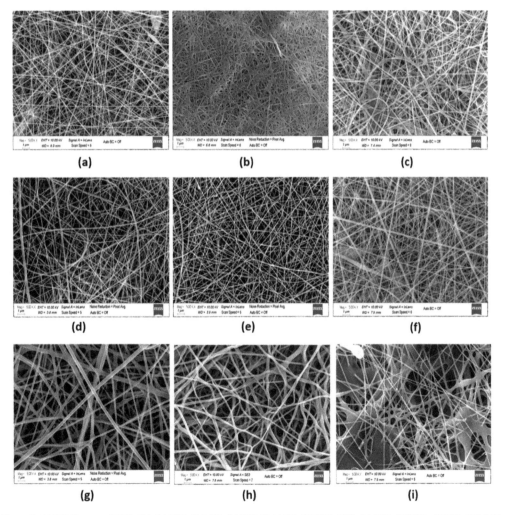

Figure 2. (**a**) SEM image of PVA-Gel, (**b**) PVA-Gel/FA, (**c**) PVA-Gel/sFA, (**d**) PVA-CHi, (**e**) PVA-CHi/FA, (**f**)PVA-CHi/sFA, (**g**) PVA-Alg, (**h**) PVA-Alg/FA, and (**i**) PVA-Alg/sFA nanofibers (Magnification: 5 kX, scale bar: 1 μm).

Polymers **2021**, *13*, 3594

Table 3. Average diameter and porosity of the fibers.

Sample	Average Diameter of Fibers (nm)	Porosity (%)
PVA-Gel	101.07 ± 30.0	46.55 ± 2.30
PVA-Gel/FA	115.23 ± 39.8	34.88 ± 4.93
PVA-Gel/sFA	145.33 ± 31.65	43.46 ± 1.16
PVA-CHi	96.26 ± 25.3	48.93 ± 0.82
PVA-CHi/FA	92.66 ± 23.9	52.09 ± 3.47
PVA-CHi/sFA	101.86 ± 36.02	42.97 ± 1.22
PVA-Alg	269.51 ± 70.25	44.31 ± 2.35
PVA-Alg/FA	208.2 ± 47.32	48.97 ± 2.66
PVA-Alg/sFA	158.17 ± 48.55	43.97 ± 7.49
PVA-Gel–CHi	96.15 ± 29	54.3 ± 3.79
PVA-Gel–CHi/FA	106.94 ± 32.13	47.15 ± 1.03
PVA-Gel–CHi/sFA	76.82 ± 20	52.2 ± 2.14
PVA-Alg–CHi	85.96 ± 30.4	52.2 ± 2.13
PVA-Alg–CHi/FA	84.45 ± 29.07	51.21 ± 0.94
PVA-Alg–CHi/sFA	123.53 ± 66.16	51.34 ± 2.01

PVA-Gel–CHi fibers have a uniform shape and are randomly oriented. As FA was incorporated into the polymer solution, the interaction of the fibers increased and a rough structure was formed. The same situation was not observed when adding FA via the electrospraying method (Figure 3c). Decreasing the viscosity due to the addition of the NaHCO$_3$ caused a different web structure. PVA-Gel–CHi/sFA fibers were almost the same as the neat fibers. These samples showed almost no fiber entanglement. Nevertheless, the average fiber diameter of the neat PVA-Gel–CHi sample decreased from 96.15 nm to 76.82 nm (Table 3). This could be because, during the e-spray process, the FA and polymer solutions dissolve in each other before reaching the collector.

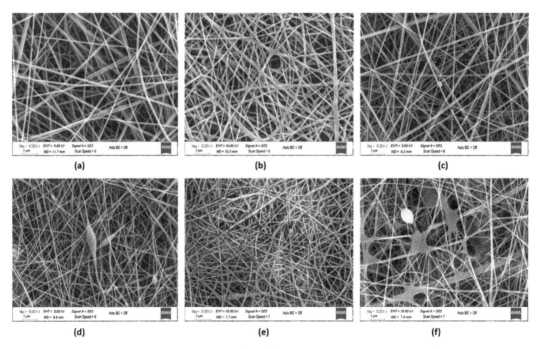

Figure 3. (a) SEM image of PVA-Gel–CHi, (b) PVA-Gel–CHi/FA, (c) PVA-Gel–CHi/sFA, (d) PVA-Alg–CHi, (e) PVA-Alg–CHi/FA, and (f) PVA-Alg–CHi/sFA nanofibers (Magnification: 10 kX, scale bar: 1 μm).

The PVA-Alg–CHi web had a bead-on-string structure and thin and continuous fibers (Figure 3d). The PVA-Alg–CHi/FA web included discontinuous fibers and had a non-uniform web morphology, but not a bead-on-string structure. The addition of FA increased the fiber interaction and fiber breakdown due to increase secondary interactions in the electrospinning process (Figure 3e) (Li et al., 2012). In the PVA-Alg–CHi/sFA web, a PVA-Alg/sFA film formed as a web. The film formation occurred for the same reasons as in PVA-Alg/sFA.

3.2. Chemical Structure of the Hybrid Nanofibers

FT-IR spectra were acquired to investigate the physical interactions between polymer matrices and FA particles, as illustrated in Figures 4–6. It was determined that FA particles were successfully sprayed onto the surface of PVA-Gel nanofibers due to the intense and clearly defined C=O stretching of folic acid at 1651 cm^{-1}. The band of the (–OH) group occurs at 3290 cm^{-1} in PVA-Gel fibers. The bands at 2938 and 2913 cm^{-1} are from the –CH$_2$ stretching vibration. The typical absorption peaks of gelatin are mostly due to peptide bonds (–CONH) with amide I–III vibrations. The peak at 1643 cm^{-1} (amide-I) is associated with –C=O stretching vibration, while the peak at 1535 cm^{-1} (amide-II) is related with N-H bending and C-H stretching vibration. The (amide-III) peak was recorded at 1243 cm^{-1}. Further, 1435 cm^{-1} (–CH$_2$ bending), 1374 cm^{-1} (C–H wagging), 1088 cm^{-1} (–C–O–C), and 837 cm^{-1} (C–C) stretching were obtained. The disappearance of specific peaks in the FT-IR spectra of PVA-Gel/FA (Figure 5a) indicates that FA was loaded into the nanofibers. It could imply that the FA is uniformly spread throughout the blended nanofiber.

The peaks at 3590 cm^{-1}, 3496 cm^{-1}, and 3330 cm^{-1} are linked with the (–OH) groups of folic acid in the crystalline structure. Moreover, the characteristic peaks of folic acid in the broad band between 3101 and 2400 cm^{-1} are caused by (–OH) groups of glutamic acid [35]. It has been found that the –NH stretching peak in the structure of the pterin ring disappeared in the broad band; however, the peak can be seen at 3101 cm^{-1} [36]. The (–C=O) group, related to pterin structure, can be observed at around 1694 cm^{-1} [35]. The peaks at 1635 cm^{-1} ((–C=N) stretching) and 1597 cm^{-1} belong to the bending of (–CONH$_2$) [37], and 1477 cm^{-1} is related to the (–C=C) stretching of phenyl and pterin rings.

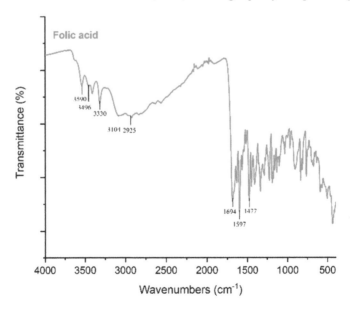

Figure 4. FT-IR spectra of folic acid.

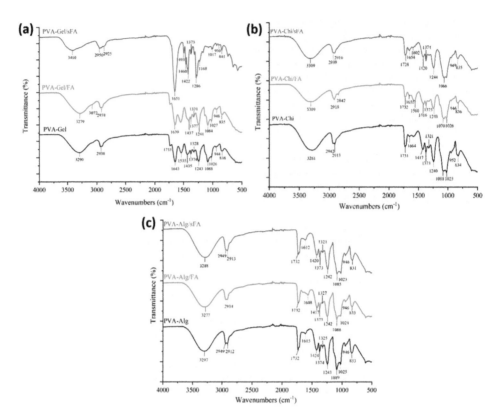

Figure 5. FT-IR spectra of binary nanofibers, (**a**) shows comparison of PVA-Gel/sFA, PVA-Gel/FA, and PVA-Gel, (**b**) shows spectra of PVA-Chi/sFA, PVA-Chi/FA, and PVA-Chi, and (**c**) shows spectra of PVA-Alg/sFA, PVA-Alg/FA, and PVA-Alg.

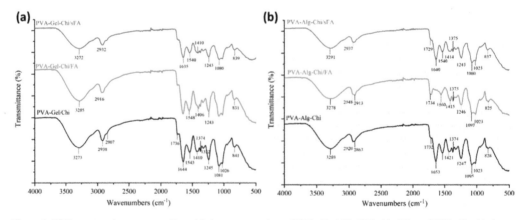

Figure 6. FT-IR spectra of ternary nanofibers (**a**) shows comparison of PVA-Alg/sFA, PVA-Alg/FA, and PVA-Alg, (**a**) shows spectra of PVA-Gel-Chi/sFA, PVA-Gel-Chi/FA, and PVA-Gel-Chi, and (**b**) shows spectra of PVA-Alg-Chi/sFA, PVA-Alg-Chi/FA, and PVA-Alg-Chi.

In PVA-CHi fibers, the band of the (–OH) group appears at about 3281 cm^{-1}, related to –OH and –NH stretching vibrations. The FTIR spectra of the PVA/chitosan fibers (Figure 6b) show characteristic peaks of PVA and chitosan, with the exception of those associated with the ionization of the major amino groups of chitosan [38]. These peaks are signals at 1417 cm^{-1} and 1560 cm^{-1}. The symmetric deformation of –NH$_3^+$ groups is responsible for the formation of the peak around 1560 cm^{-1}, while the peak at 1417 cm^{-1} refers to carboxylic acid and those at 1321 cm^{-1} and 1240 cm^{-1} are connected to C-H vibration [39]. The peaks at 1731 cm^{-1} are typical of carboxylic acid dimers. The results are consistent with the literature [40].

The typical peaks of the PVA-Alg nanofiber were identified at 3297 cm^{-1}, 2949 cm^{-1}, 2912 cm^{-1}, 1732 cm^{-1}, and 1089 cm^{-1}, attributed to –OH groups, asymmetric and symmetric –CH stretching, and –CO stretching, respectively (Figure 5c). The characteristic peaks of the electrospun PVA-Alg/FA nanofibers were detected at 3277 cm^{-1}, 2949 cm^{-1}, 2914 cm^{-1}, 1732 cm^{-1} and 1086 cm^{-1}, and are related to –OH groups, asymmetric and symmetric –CH stretching, and CO stretching, respectively. Moreover, the peaks at 1615 cm^{-1} and 1424 cm^{-1}, resulting from asymmetric and symmetric (–COO) stretching, in the PVA-Alg fibers are slightly shifted to 1608 cm^{-1} and 1417 cm^{-1} in the PVA-Alg/FA fibers.

The characteristic peak of the carboxyl group of alginate, the amide group of chitosan, and the overlapping PVA peaks are similar to those in the PVA-Gel–CHi and PVA-Alg–CHi nanofibers (Figure 6a,b).

3.3. Thermal Properties and Weight-Loss of Nanofibers and Folic Acid

Figure 7 shows the thermogram of folic acid, which reveals that the bioactive powder was decomposed into four stages. In the first stage, the weight-loss of 7.54% was caused by removing moisture at temperatures from 100 to 170 °C. The 2.34% weight-loss in the second stage is related to the degradation of glutamic acid in the FA structure (between 170 and 240 °C) [41]. A slower degradation behavior occurred in the 240–600 °C range, with a weight-loss of about 49%. This was due to the degradation of the functional groups, which are pterin and p-amino benzoic acid (Janković, 2010). In the last stage, the 41.3% weight-loss was connected to the decomposition of pyrolysis products formed in the second and third steps.

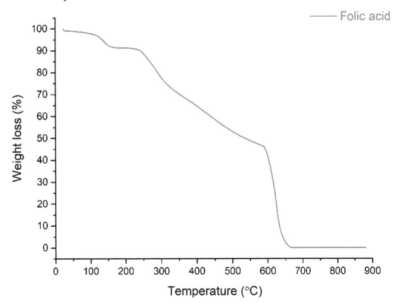

Figure 7. TGA thermogram of folic acid.

TGA was used to assess the thermal stability of the nanofibers (Figure 8). Three main stages are seen in the TGA curves, which are linked to solvent and humidity being trapped inside (30–110 °C) the nanofiber samples; the second stage is connected to degradation of the polymers between 280 °C and 450 °C and the third stage is due to the degradation of the pyrolysis products of the polymers, which are formed during the polymer degradation stage under a N_2 atmosphere [10,42]. The second weight-loss phase, related to the decomposition of the polymers, was composed of two steps according to the binary polymer mixture. The decomposition curves of PVA-Gel, PVA-Alg and PVA-CHi suggest two different decomposition behaviors. The first decomposition was related to the removal of the side group from the macromolecule chain. The last was related to the main macromolecule chain scission. Except for PVA-Alg/FA and PVA-Alg/FA, all samples gave pyrolysis products due to the inert atmosphere (N_2 condition). After changing the atmosphere from N_2 to O_2 at 600 °C, all samples lost mass, except PVA-Alg/FA and PVA-Alg/FA, resulting from the alginate decomposition behavior in the inert atmosphere. Alginate does not create pyrolysis products when it decomposes under inert condition [43].

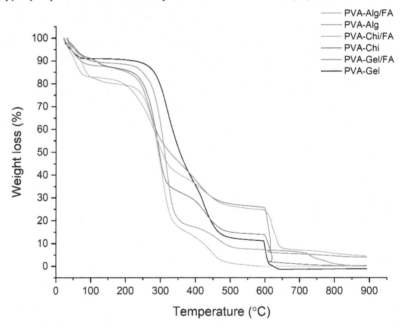

Figure 8. TGA thermogram of nanofibers.

The decomposition of pure PVA has been discussed in the literature [31]. The thermal decomposition of the PVA mixture was hardly different from that of the pure PVA polymer, especially the last part of the curve. The first part of the polymer decomposition curve was also changed by adding FA molecules into the polymers. FA addition decreases the polymer decomposition speed due to its protective properties. This situation implies that FA could be used as an anti-oxidant in protective polymers during processing at high temperatures. FA's conservation properties did not emerge in the case of the alginate polymers. The decomposition speed of the alginate mixture was not changed by adding FA due to the behavior of alginate decomposition in the inert atmosphere. The beginning of the first polymer decomposition stage was begun at lower temperatures by adding FA molecules. This resulted from the decomposition behavior of the FA materials [10,44]. The early decomposition of the FA molecules caused energy absorption, similar to the antioxidant running mechanism [45]. This energy absorption process preserves and delays the main polymers' decomposition.

3.4. In Vitro Release Test and Kinetics

UV-Vis spectroscopy was utilized to analyze the FA release profile of the nanofibers, which assisted in revealing the structure–function connection of the electrospun fibers containing FA in the artificial acid (pH 5.44, 37 °C) and alkaline (pH 8.04, 37 °C) sweat solutions over 480 min. Figures 9 and 10 depict the cumulative FA release rate profiles in the sweat solution samples, which were binary and ternary nanofibers containing FA. In Figure 9a, the release profiles of the PVA-Gel-sFA nanofibers in both acidic and alkaline media are similar, and they were released quickly in the first 30 min, then these samples continued to release FA slowly until 480 min. After 210 min, the FA release rate in alkaline media increased compared to acidic medium, and the maximum release value increased to almost 95%. The release behavior of the PVA-Gel/FA samples (in both acidic and alkaline medium) was partially different from that of PVA-Gel/sFA samples. These samples showed fast release behavior in alkaline medium for the first 30 min, while the fast release behavior in acidic medium continued until 90 min. Moreover, the amount of FA released in the alkaline medium was higher than in the acidic medium until 360 min. After this, the release rate increased in the acidic medium, and the cumulative FA release reached about 90%. It has been reported that the release media affects the polymer matrix, as well as the active drug [46].

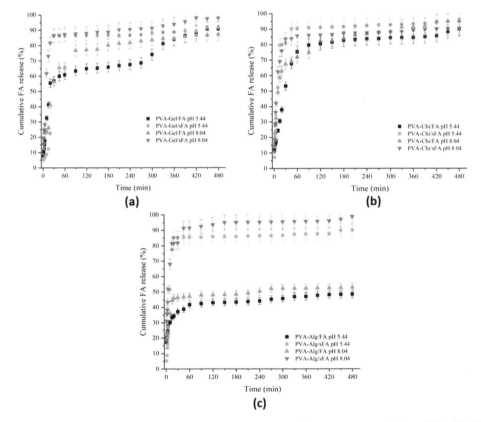

Figure 9. UV-Vis release of binary nanofibers (**a**) release behavior of PVA-Gel/FA at pH 5.44, and 8.04, and PVA-Gel/sFA at pH 5.44 and 8.04, (**b**) release behavior of PVA-Chi/FA at pH 5.44 and 8.04, and PVA-Chi/sFA at pH 5.44 and 8.04, while (**c**) shows release behavior of PVA-Alg/FA at pH 5.44 and 8.04, and PVA-Alg/sFA at pH 5.44 and 8.04 respectievly.

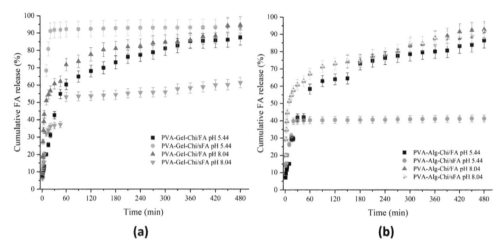

Figure 10. UV-Vis release of ternary nanofibers (**a**) shows release behavior of PVA-Gel-Chi/FA at pH 5.44 and 8.04, and PVA-Gel-Chi/sFA at pH 5.44 and 8.04, while (**b**) shows release behavior of PVA-Alg-Chi/FA at pH 5.44 and 8.04, and PVA-Alg-Chi/sFA at pH 5.44 and 8.04.

It could be said that the fast release of FA from the samples produced via the electro-spraying process depended on the physical bond between the FA molecules and the fiber surface. It can be seen from Figure 2c that FA molecules were deposited on the fiber surface. When this type of web is immersed in acidic or alkaline media, the media breaks down the weak physical bond between fiber and FA molecules. The fast release of the samples in which FA was embedded was related to the depositing of FA on the fiber surface. Because of this, the release ratio was lower than that in the other samples produced via electrospraying. During the controlled release period, the release speed of the samples produced via electrospinning was higher than that of the others due to the removing of the embedded FA from inside the fiber.

The release behaviors of PVA-CHi/FA samples were the same in both acidic and alkaline medium (Figure 9b). After 180 min, the FA release rate increased in alkaline medium, and therefore the samples showed the highest quantity of released FA at the end of the release study. The amount released in acidic medium remained at almost 90%. Until 90 min, the release behavior of PVA-CHi/sFA was different from that of PVA-CHi/FA. A higher FA release rate occurred in the acidic medium (around 93%), while a lower FA release rate was observed in alkaline medium (around 85%).

Nair et al. (2019) performed a release study of curcumin-loaded chitosan nanoparticles by simulating skin at two distinct pH values (pH 5 and pH 7.4) [47]. In this study, the change in curcumin release observed under two different pH conditions was associated with the pH-dependent swelling behavior of chitosan. The swelling of chitosan allows the release medium to penetrate the polymer matrix and act as a plasticizer, which transforms the glassy polymer into a more rubbery structure, leading to increased drug release from the nanoparticle structure [48]. The release profiles of PVA-CHi/FA samples in both acidic and alkaline media are similar. Similarly, the release profiles of PVA-Kit/sFA in acidic and alkaline media are similar to each other. The PVA-CHi/FA samples released FA at 90% and 95% in acidic and alkaline media, respectively, due to the dissolution of PVA and chitosan. Mafenite acetate-loaded PVA-chitosan nanofibers were found to release 50% of the drug in the web structure within the first 10 min in the PBS medium [49].

The release rates of FA from the PVA-Alg/sFA nanofibers were similar in acidic and alkaline media (Figure 9c). PVA-Alg/sFA samples showed a fast release rate in the first 60 min, and then controlled release after this. It can be observed that the amount of FA released in these media during the analysis was over 90%. These findings are in agreement with the literature. Lutein-loaded sodium alginate/PVA nanofibers were found to achieve very fast lutein release, and cumulative release was completed in 7 min, at almost 92% [20]. The fast release in the first few minutes is explained by the hydrophilic character of alginate and the PVA composition. Arthanari et al. (2016) noted that gatifloxacin-loaded PVA-sodium alginate nanolyphs did not display a fast burst release tendency in the PBS (pH 7.4) for the first 1 h, but gatifloxacin was released (95% in total) during the test period [50]. The PVA-Alg/FA nanofibers in acidic and alkaline media were also similar, with almost 50% of FA released. PVA-Alg/FA had a fast release rate in the first 60 min. Unlike PVA-Alg/sFA samples, the PVA-Alg/FA samples' FA release rate increased after 240 min. The PVA-Alg samples had the same release behavior as in PVA-Gel.

As seen Figure 10a, the ternary mixture of the PVA-Gel–CHi/FA sample had the same release profile in both acidic and alkaline media. These samples displayed fast release in the first 60 min. After this time, the release rate increased in a controlled manner. The release behaviors of PVA-Gel–CHi/sFA samples differed from those of PVA-Gel–CHi/FA samples. In addition, the release behaviors of PVA-Gel–CHi/sFA samples in acidic and alkaline media vary from each other. These samples showed fast release rates in alkaline medium in the first 60 min, and the release amount reached 60% in a controlled manner. In acidic medium, the samples showed fast release in the first 30 min and this was sustained in a controlled manner.

Apart from the PVA-Alg–CHi/sFA nanofiber structure in the acidic medium, the release rates of the other three webs were similar in both acidic and alkaline media (Figure 10b). For PVA-Alg–CHi/sFA nanofibers, the FA released in acidic medium was considerably low compared to the alkaline medium (41%). This could be related to the homogeneity of the electrospraying process. Furthermore, PVA-Alg–CHi/sFA exhibited a controlled release behavior after 30 min, while the other three samples showed gradual release to almost 90%. The PVA-Alg–CHi/FA samples showed little difference compared to the other two samples in the alkaline media.

Crosslinkers are often utilized to prepare PVA and/or gelatin-based scaffolds with physicochemical characteristics, and the obtained scaffolds are almost identical to healthy tissues. Besides this, drug release rates are advanced by crosslinkers. However, crosslinking reactions minimize the bio-based scaffolds' in vivo degradation, and negatively affect host tissue responses [51,52]. In particular, gluteraldehyde (GA), which can be used as a crosslinker, can be considerably toxic to the body, and is typically restricted in many medicinal applications [53].

In the decision regarding the release mechanism, the choice of nanofiber matrix plays a critical role, dependent on degradation behavior and interaction with theurapeutics [54]. The diffusion mechanisms are based on both the polymer matrix properties and the processing variables, that is, whether they are Fickian or non-Fickian [55]. On the other hand, the solution matrix is more suitable for therapeutics, as these are dispersed in the electrospun fibers, resulting from the release rate and fiber morphology at which Fickian diffusion predominates. If a matrix shows considerably slow drug release, the Higuchi model (Power Law) is most often used to explain the diffusion-driven release. However, polymer degradation usually plays a substantial role in drug release, as is the case with PVA, in addition to release due to diffusion, when the matrix is biodegradable. The Korsmeyer–Peppas kinetic model typically defines such mixed release mechanisms [54].

The release profiles of the different nanofibers in two distinct release media correlated the best with different mathematical models (Table S1). The PVA-Alg–CHi/sFA nanofibers (in pH 5.44) gave the best results with zero-order, first-order, and Hixson–Crowell kinetic models. Generally, the Korsmeyer–Peppas kinetic model is used more than other models (Table S1). Moreover, the "n exponential factor" was found to be $n < 1$ and $0.45 < n < 0.89$ for the nanofibers. For matrices in cylindrical form, the Fickian diffusion mechanism and signs of non-Fickian or abnormal transport are evident at $0.45 < n < 0.89$. It was observed that both the change of release media and the use of different polymer matrices affects the release mechanism. It was found that the change in pH in the mechanism of FA release from the nanofibers obtained by the e-spraying method did not have a significant effect on the "n exponential factor". However, the release of FA from nanofibers obtained by electrospinning was partially changed, depending on pH.

3.5. Entrapment Efficiency (EE, %) and Loading Capacity (LC, %)

The entrapment efficiency (EE, %) and also loading capacity (LC, %) have been calculated in the manuscript [56]. These values have been determined by the non-volatile FA structure using a UV-Vis spectrophotometer. The drug solubility of electrospun fibers influences the entrapment efficiency and the loading capacity of FA-loaded and FA-sprayed nanofibers (Table S2).

3.6. Cytotoxic Effects of Nanofibers by MTT Assay

The hydrophilic character of these composite fibers, which may assist in connecting the desired cells to the surfaces of the scaffolding, thus enables cell development and the improvement of viability, which is desirable in the context of applications of tissue engineering [57]. The proliferation of cells seeded on electrospun FA-loaded and FA-sprayed composite and pure nanofibers was evaluated via MTT assay after 24 h of treatment in L929 cells. The MTT results indicate that not all nanofiber scaffolds showed considerable toxicity, since their viability was no less than 70% (Figure 11) [34]. All PVA-based composite fibers with FA could be widely utilized in wound dressing and tissue engineering fields due to their ability to support cell function. The results prove that the addition of FA into PVA-based nanofibers enhances cell growth and supports cell proliferation. FA-loaded bicomponent composite fibers promoted cell proliferation during the MTT assays. In this regard, the blending process is more efficient than simultaneous processes. This could be related to the excess alcohol from the FA solution affecting the nanofibers.

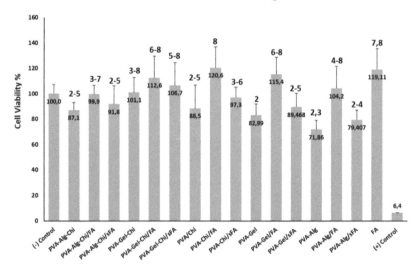

Figure 11. Effects of all nanofibers on cell viability of L929 cells, by MTT assay.

The PVA-Gel/FA, PVA-CHi/FA, and PVA-Alg/FA samples acheved cell viability values of 115.4 ± 13.1, 120.6 ± 16.4, and 104.2 ± 17, respectively. PVA-CHi/FA fibers had the highest cell viability value of 120.6 ± 16.4.

Folic acid plays a crucial role in cell growth and repair, assisting wound healing via the biosynthesis of nucleic acids [58,59]. It is pointed out that FA can be converted into active tetrahydrofolic acid owing to the response of folic acid reductase in vivo. As a major provider of a carbon unit, tetrahydrofolic acid plays a key role in purine and pyrimidine synthesis and in amino acid mutual transformation; the absence of this in vivo vitamin may affect the transmission of the carbide unit and may affect nucleic acid synthesis bad amino acid metabolism, which are necessary to proliferate, grow and develop cells. The presence of FA in cell division throughout the wound healing process improves the healing rate due to its effect on growing cell division [60].

All the electrospun composite fibers consisting of FA exhibited more cell viability (%) compared to pure ones. This can be explained by the fact that FA has a high affinity for the folate receptor on the L929 cells [38]. Furthermore, PVA-based nanofibers have been noted as excellent drug carriers in a matrix, due to their non-toxic and biocompatible features. Therefore, PVA is utilized for a large range of medicinal applications, especially in its fibrous form, through these characteristics [61–63].

The ECM structure of fibroblast, osteoblast, chondrocyte and endothelial cells are the same in micro and nanofibers arranged in a random fibrous mat, which allows for quick regeneration and differentiation [64]. In recent years, various wound dressings have been developed and evaluated with similar cytotoxicity results to ours. Kalalinia et al. (2021) developed vancomycin (VCM)-loaded hybrid chitosan/poly ethylene oxide (Chi/PEO) nanofibers by blend-electrospinning [65]. There are no harmful effects of any group, according to the findings of the Alamar Blue cytotoxicity tests performed on HDFs. Çetmi et al. (2019) fabricated microalgal extract-loaded PCL nanofibers, and MTT cell growth assays revealed that the new fibers enhance the growing of cardiovascular cells by contributing significantly to cell proliferation [66]. Similarly, carbon nanofiber/gold nanoparticle (CNF/GNP)-conductive nanofibers used as scaffold have been produced by Nekounam et al. (2020) using two strategies: e-spraying simultaneously with electrospinning and blending electrospinning methods [67]. In this study, the MTT assays performed on MG63 cells did not show any toxicity, and all obtained nanofibers showed biocompatibility with cells.

These PVA-based nanofibers not only exhibited good biocompatibility in vitro, but were also found to possess many preferable wound/tissue engineering applications. In sum, PVA-based FA scaffolds offered the best nanofibers for cell development and are appropriate candidates for tissue engineering and wound healing.

According to the SNK analysis, all nanofibers ranked between the positive control (1) and PVA-CHi/FA (8), the cell viability of which was higher than FA. FA promotes the growth of cells because its cell viability values are greater than negative. PVA-Gel and PVA-Alg nanofibers are in the same group (2). The PVA-Alg sample is also placed in the third group. PVA-Alg–CHi, PVA-CHi, PVA-Gel/sFA and PVA-Alg–CHi/sFA nanofibers are in the same groups (2, 3, 4, and 5). Moreover, PVA-Alg–CHi/FA are in the same groups (3, 4, 5, 6, and 7) as the negative control. The PVA-CHi/FA, PVA-Gel/FA, PVA-Gel–CHi/FA and PVA-Gel–CHi/Sfa samples are considered to be more effective in terms of cell viability than others, based on the SNK analysis. The MTT assay results of all nanofibers show they would be acceptable for use as biocompatible materials in accordance with ISO 10993-5.

4. Conclusions

In the current study, FA-loaded and FA-sprayed PVA-based nanofibers were successfully fabricated via two distinct approaches: electrospinning and simultaneous electrospinning and electrospraying processes. The SEM micrographs indicate that spraying FA onto web layers and/or surfaces affects the nanofiber's surface morphology. On the other hand, FA-loading did not change the morphology without phase separation, owing to the compatibility of FA with polymer matrices. The PVA gel kit samples achieved the highest porosity of $54.3 \pm 3.8\%$, while PVA gel kit/sFA samples had the lowest fiber diameter (76.8 ± 2 nm) based on ImageJ software measurement. Based on FT-IR analysis, the resultant FA-sprayed composite fibers included the characteristic peaks of both polymer and FA. Furthermore, the TGA thermogram confirmed FA can be used as an antioxidant agent to delay the decomposition of polymer blends in the processing stage. PVA-CHi/FA webs had the highest release value of 95.2% in alkaline media, whereas PVA-Gel–CHi/sFA webs released the most FA (92%) in acidic media. The highest entrapment efficiency value was determined as 94.82 ± 4.7% for the PVA-Gel/sFA sample, while the lowest entrapment efficiency was determined for PVA-Alg/FA at 50.66 ± 2.5%. The loading capacity values of FA-loaded and FA-sprayed nanofibers were lower than 3%. Meanwhile, the MTT assays of PVA-CHi/FA samples showed that they had the highest cell viability, at 120.6 ± 16.4. Moreover, the FA release mechanism from the nanofibers was either Fickian or non-Fickian (abnormal transport) due to the change in matrices. The cytotoxicity studies of L929 cells performed via MTT assay displayed good biocompatibility, and even these hybrid nanofibers supported cell proliferation. In view of the results, nanofibrous PVA-based materials containing FA samples are recommended for tissue scaffold applications.

Supplementary Materials: The following are available online at https://www.mdpi.com/article/10.3390/polym13203594/s1. Table S1: Regression values of the FA-loaded and FA-sprayed nanofibers, Table S2: The entrapment efficiency (%) and loading capacities (%) of all nanofibers.

Author Contributions: Conceptualization, F.N.P.; methodology, F.N.P. and M.H.; software, S.U.; validation, K.Y. and I.-S.K.; formal analysis, F.N.P. and S.U.; investigation, F.N.P.; resources, F.N.P. and I.-S.K.; data curation, S.U. and M.H.; writing—original draft preparation, F.N.P. and K.Y.; writing—review and editing, S.U. and K.Y.; visualization, I.-S.K.; supervision, F.N.P. and I.-S.K.; project administration, K.Y.; funding acquisition, F.N.P. and I.-S.K. All authors have read and agreed to the published version of the manuscript.

Funding: This study was funded by the Scientific Research Unit of the Bursa University of Technology (SRP Project No. 190D003).

Institutional Review Board Statement: Not applicable.

Informed Consent Statement: Not applicable.

Data Availability Statement: The data presented in this study are available on request from the corresponding author.

Acknowledgments: This study was funded by the Scientific Research Unit of the Bursa University of Technology (SRP Project No. 190D003). The authors would like to thank Bursa Technical University Central Research Laboratory (Bursa, Turkey) for the SEM and TGA analysis and Gökçe Taner for MTT assays. The authors would like to thank as well the employees of the Scientific and Technological Research Council of Turkey—Bursa Test and Analysis Laboratory (TUBİTAK-BUTAL, Turkey) for supporting the preparation of the artificial sweat solutions.

Conflicts of Interest: The author(s) declare no potential conflict of interest with respect to the research, authorship, and/or publication of this article.

References

1. Gupta, K.C.; Haider, A.; Choi, Y.-R.; Kang, I.-K. Nanofibrous scaffolds in biomedical applications. *Biomater. Res.* **2014**, *18*, 1–11. [CrossRef]
2. Hashmi, M.; Ullah, S.; Ullah, A.; Akmal, M.; Saito, Y.; Hussain, N.; Ren, X.; Kim, I.S. Optimized Loading of Carboxymethyl Cellulose (CMC) in Tri-component Electrospun Nanofibers Having Uniform Morphology. *Polymers* **2020**, *12*, 2524. [CrossRef]
3. Ullah, S.; Ullah, A.; Lee, J.; Jeong, Y.; Hashmi, M.; Zhu, C.; Joo, K.I.; Cha, H.J.; Kim, I.S. Reusability Comparison of Melt-Blown vs Nanofiber Face Mask Filters for Use in the Coronavirus Pandemic. *ACS Appl. Nano Mater.* **2020**, *3*, 7231–7241. [CrossRef]
4. Ullah, S.; Hashmi, M.; Hussain, N.; Ullah, A.; Sarwar, M.N.; Saito, Y.; Kim, S.H.; Kim, I.S. Stabilized nanofibers of polyvinyl alcohol (PVA) crosslinked by unique method for efficient removal of heavy metal ions. *J. Water Process Eng.* **2020**, *33*, 101111. [CrossRef]
5. Illangakoon, U.E.; Gill, H.; Shearman, G.C.; Parhizkar, M.; Mahalingam, S.; Chatterton, N.P.; Williams, G.R. Fast dissolving paracetamol/caffeine nanofibers prepared by electrospinning. *Int. J. Pharm.* **2014**, *477*, 369–379. [CrossRef] [PubMed]
6. Li, X.; Kanjwal, M.A.; Lin, L.; Chronakis, I.S. Electrospun polyvinyl-alcohol nanofibers as oral fast-dissolving delivery system of caffeine and riboflavin. *Colloids Surf. B Biointerfaces* **2013**, *103*, 182–188. [CrossRef] [PubMed]
7. Parın, F.N.; Yıldırım, K. Preparation and characterisation of vitamin-loaded electrospun nanofibres as promising transdermal patches. *Fibres Text. East. Eur.* **2021**, *29*, 17–25. [CrossRef]
8. Lu, Y.; Huang, J.; Yu, G.; Cardenas, R.; Wei, S.; Wujcik, E.K.; Guo, Z. Coaxial electrospun fibers: Applications in drug delivery and tissue engineering. *Wiley Interdiscip. Rev. Nanomed. Nanobiotechnol.* **2016**, *8*, 654–677. [CrossRef]
9. Li, X.-Y.; Li, Y.-C.; Yu, D.-G.; Liao, Y.-Z.; Wang, X. Fast disintegrating quercetin-loaded drug delivery systems fabricated using coaxial electrospinning. *Int. J. Mol. Sci.* **2013**, *14*, 21647–21659. [CrossRef]

10. Parın, F.N.; Aydemir, Ç.İ.; Taner, G.; Yıldırım, K. Co-electrospun-electrosprayed PVA/folic acid nanofibers for transdermal drug delivery: Preparation, characterization, and in vitro cytocompatibility. *J. Ind. Text.* **2021**, 1528083721997185. [CrossRef]
11. Hashmi, M.; Ullah, S.; Kim, I.S. Copper oxide (CuO) loaded polyacrylonitrile (PAN) nanofiber membranes for antimicrobial breath mask applications. *Curr. Res. Biotechnol.* **2019**, *1*, 1–10. [CrossRef]
12. Kharaghani, D.; Gitigard, P.; Ohtani, H.; Kim, K.O.; Ullah, S.; Saito, Y.; Khan, M.Q.; Kim, I.S. Design and characterization of dual drug delivery based on in-situ assembled PVA/PAN core-shell nanofibers for wound dressing application. *Sci. Rep.* **2019**, *9*, 1–11. [CrossRef]
13. Parin, F.N.; Terzioğlu, P.; Sicak, Y.; Yildirim, K.; Öztürk, M. Pine honey–loaded electrospun poly (vinyl alcohol)/gelatin nanofibers with antioxidant properties. *J. Text. Inst.* **2021**, *112*, 628–635. [CrossRef]
14. Ullah, A.; Ullah, S.; Khan, M.Q.; Hashmi, M.; Nam, P.D.; Kato, Y.; Tamada, Y.; Kim, I.S. Manuka honey incorporated cellulose acetate nanofibrous mats: Fabrication and in vitro evaluation as a potential wound dressing. *Int. J. Biol. Macromol.* **2020**, *155*, 479–489. [CrossRef]
15. Gunn, J.; Zhang, M. Polyblend nanofibers for biomedical applications: Perspectives and challenges. *Trends Biotechnol.* **2010**, *28*, 189–197. [CrossRef]
16. Ullah, S.; Hashmi, M.; Kharaghani, D.; Khan, M.Q.; Saito, Y.; Yamamoto, T.; Lee, J.; Kim, I.S. Antibacterial properties of in situ and surface functionalized impregnation of silver sulfadiazine in polyacrylonitrile nanofiber mats. *Int. J. Nanom.* **2019**, *14*, 2693–2703. [CrossRef]
17. Ullah, S.; Hashmi, M.; Khan, M.Q.; Kharaghani, D.; Saito, Y.; Yamamoto, T.; Kim, I.S. Silver sulfadiazine loaded zein nanofiber mats as a novel wound dressing. *RSC Adv.* **2019**, *9*, 268–277. [CrossRef]
18. Materials, N.N.; Aktürk, A. Fabrication and Characterization of Polyvinyl Alcohol / Gelatin / Silver Nanoparticles Nanocomposite Materials. *Eurasian J. Biol. Chem. Sci.* **2019**, *2*, 1–6.
19. Yang, D.; Li, Y.; Nie, J. Preparation of gelatin/PVA nanofibers and their potential application in controlled release of drugs. *Carbohydr. Polym.* **2007**, *69*, 538–543. [CrossRef]
20. Han, X.; Huo, P.; Ding, Z.; Kumar, P.; Liu, B. Preparation of lutein-loaded PVA/sodium alginate nanofibers and investigation of its release behavior. *Pharmaceutics* **2019**, *11*, 449. [CrossRef] [PubMed]
21. Najafiasl, M.; Osfouri, S.; Azin, R.; Zaeri, S. Alginate-based electrospun core/shell nanofibers containing dexpanthenol: A good candidate for wound dressing. *J. Drug Deliv. Sci. Technol.* **2020**, *57*, 101708. [CrossRef]
22. Cho, D.; Netravali, A.N.; Joo, Y.L. Mechanical properties and biodegradability of electrospun soy protein Isolate/PVA hybrid nanofibers. *Polym. Degrad. Stab.* **2012**, *97*, 747–754. [CrossRef]
23. Cui, Z.; Zheng, Z.; Lin, L.; Si, J.; Wang, Q.; Peng, X.; Chen, W. Electrospinning and crosslinking of polyvinyl alcohol/chitosan composite nanofiber for transdermal drug delivery. *Adv. Polym. Technol.* **2018**, *37*, 1917–1928. [CrossRef]
24. Liu, Q.; Ouyang, W.-C.; Zhou, X.-H.; Jin, T.; Wu, Z.-W. Antibacterial Activity and Drug Loading of Moxifloxacin-Loaded Poly(Vinyl Alcohol)/Chitosan Electrospun Nanofibers. *Front. Mater.* **2021**, *8*, 1–9. [CrossRef]
25. Ghorani, B.; Tucker, N. Fundamentals of electrospinning as a novel delivery vehicle for bioactive compounds in food nanotechnology. *Food Hydrocoll.* **2015**, *51*, 227–240. [CrossRef]
26. Taheri, A.; Jafari, S.M. Gum-based nanocarriers for the protection and delivery of food bioactive compounds. *Adv. Colloid Interface Sci.* **2019**, *269*, 277–295. [CrossRef]
27. Fernández-Villa, D.; Gómez-Lavín, M.J.; Abradelo, C.; Román, J.S.; Rojo, L. Tissue engineering therapies based on folic acid and other vitamin B derivatives. Functional mechanisms and current applications in regenerative medicine. *Int. J. Mol. Sci.* **2018**, *19*, 4068. [CrossRef]
28. Alborzi, S.; Lim, L.T.; Kakuda, Y. Release of folic acid from sodium alginate-pectin-poly(ethylene oxide) electrospun fibers under invitro conditions. *LWT Food Sci. Technol.* **2014**, *59*, 383–388. [CrossRef]
29. Fonseca, L.M.; Crizel, R.L.; da Silva, F.T.; Fontes, M.R.V.; da Rosa Zavareze, E.; Dias, A.R.G. Starch nanofibers as vehicles for folic acid supplementation: Thermal treatment, UVA irradiation and in vitro simulation of digestion. *J. Sci. Food Agric.* **2021**, *101*, 1935–1943. [CrossRef]
30. Li, H.; Gao, J.; Shang, Y.; Hua, Y.; Ye, M.; Yang, Z.; Ou, C.; Chen, M. Folic Acid Derived Hydrogel Enhances the Survival and Promotes Therapeutic Efficacy of iPS Cells for Acute Myocardial Infarction. *ACS Appl. Mater. Interfaces* **2018**, *10*, 24459–24468. [CrossRef]
31. International Organization for Standardization. *ISO 105-E04:2013. Textiles-Tests for Colourfastness—Part E04: Colourfastness to Perspiration*; International Organization for Standardization: Geneva, Switzerland, 2013.
32. Moydeen, A.M.; Padusha, M.S.A.; Aboelfetoh, E.F.; Al-Deyab, S.S.; El-Newehy, M.H. Fabrication of electrospun poly(vinyl alcohol)/dextran nanofibers via emulsion process as drug delivery system: Kinetics and in vitro release study. *Int. J. Biol. Macromol.* **2018**, *116*, 1250–1259. [CrossRef]
33. International Organization for Standardization. *UNI EN ISO 10993-12:2009. Biological Evaluation of Medical Devices—Part 12: Preparation of Samples and Reference Materials*; International Organization for Standardization: Geneva, Switzerland, 2009.
34. International Organization for Standardization. *UNI EN ISO 10993-5:2009. Biological Evaluation of Medical Devices—Part 5: In Vitro Cytotoxicity Testing*; International Organization for Standardization: Geneva, Switzerland, 2009.
35. He, Y.-Y.; Wang, X.-C.; Jin, P.-K.; Zhao, B.; Fan, X. Complexation of anthracene with folic acid studied by FTIR and UV spectroscopies. *Spectrochim. Acta Part A Mol. Biomol. Spectrosc.* **2009**, *72*, 876–879. [CrossRef]

36. İnce, İ.; Yıldırım, Y.; Güler, G.; Medine, E.İ.; Ballıca, G.; Kuşdemir, B.C.; Göker, E. Synthesis and characterization of folic acid-chitosan nanoparticles loaded with thymoquinone to target ovarian cancer cells. *J. Radioanal. Nucl. Chem.* **2020**, *324*, 71–85. [CrossRef]

37. Raouf, L.A.M.; Hammud, K.; Mohammed, J.M.; Mohammed, E.K.A.-D. Qualitative and Quantitative Determination of Folic acid in Tablets by FTIR Spectroscopy. *Int. J. Adv. Pharm. Biol. Chem.* **2014**, *3*, 773–780.

38. Alhosseini, S.N.; Moztarzadeh, F.; Mozafari, M.; Asgari, S.; Dodel, M.; Samadikuchaksaraei, A.; Kargozar, S.; Jalali, N. Synthesis and characterization of electrospun polyvinyl alcohol nanofibrous scaffolds modified by blending with chitosan for neural tissue engineering. *Int. J. Nanomed.* **2012**, *7*, 25–34.

39. Ali, M.; Gherissi, A. Synthesis and characterization of the composite material PVA/Chitosan/5% sorbitol with different ratio of chitosan. *Int. J. Mech. Mechatron. Eng.* **2017**, *17*, 15–28.

40. Pakolpakçıl, A.; Draczynski, Z. Green approach to develop bee pollen-loaded alginate based nanofibrous mat. *Materials* **2021**, *14*, 2775. [CrossRef] [PubMed]

41. Jankovi, B. Thermal stability investigation and the kinetic study of Folnak® degradation process under nonisothermal conditions. *AAPS PharmSciTech* **2010**, *11*, 103–112. [CrossRef] [PubMed]

42. Shojaie, S.; Rostamian, M.; Samadi, A.; Alvani, M.A.S.; Khonakdar, H.A.; Goodarzi, V.; Zarrintaj, R.; Servatan, M.; Asefnejad, A.; Baheiraei, N.; et al. Electrospun electroactive nanofibers of gelatin-oligoaniline/Poly (vinyl alcohol) templates for architecting of cardiac tissue with on-demand drug release. *Polym. Adv. Technol.* **2019**, *30*, 1473–1483. [CrossRef]

43. Shamshina, J.L.; Gurau, G.; Block, L.E.; Hansen, L.K.; Dingee, C.; Walters, A.; Rogers, R.D. Chitin–calcium alginate composite fibers for wound care dressings spun from ionic liquid solution. *J. Mater. Chem. B* **2014**, *2*, 3924–3936. [CrossRef]

44. Parın, F.N. Yaşlanma Geciktirici B9 Vitamini İçeren Farklı Non-Woven Kumaşların Üretimi ve Kontrollü Salımının İncelenmesi. Ph.D. Thesis, Fen Bilimleri Enstitüsü, Bursa Teknik Üniversitesi, Bursa, Turkey, 2021.

45. Atteia, B.M.R.; El-Kak, A.E.-A.A.; Lucchesi, P.A.; Delafontane, P. Antioxidant activity of folic acid: From mechanism of action to clinical application. *FASEB J.* **2009**, *23*, 103.7. [CrossRef]

46. Ouazib, F.; Mokhnachi, N.B.; Haddadine, N.; Barille, R. Role of polymer/polymer and polymer/drug specific interactions in drug delivery systems. *J. Polym. Eng.* **2019**, *39*, 534–544. [CrossRef]

47. Nair, R.S.; Morris, A.; Billa, N.; Leong, C.O. An Evaluation of Curcumin-Encapsulated Chitosan Nanoparticles for Transdermal Delivery. *AAPS PharmSciTech* **2019**, *20*, 1–13. [CrossRef] [PubMed]

48. Spinks, G.M.; Lee, C.K.; Wallace, G.G.; Kim, S.I.; Kim, S.J. Swelling behavior of chitosan hydrogels in ionic liquid-water binary systems. *Langmuir* **2006**, *22*, 9375–9379. [CrossRef] [PubMed]

49. Abbaspour, M.; Makhmalzadeh, B.S.; Rezaee, B.; Shoja, S.; Ahangari, Z. Evaluation of the antimicrobial effect of chitosan/polyvinyl alcohol electrospun nanofibers containing mafenide acetate. *Jundishapur J. Microbiol.* **2015**, *8*, e24239. [CrossRef] [PubMed]

50. Arthanari, S.; Mani, G.; Jang, J.H.; Choi, J.O.; Cho, Y.H.; Lee, J.H.; Cha, S.E.; Oh, H.S.; Kwon, D.H.; Jang, H.T. Preparation and characterization of gatifloxacin-loaded alginate/poly (vinyl alcohol) electrospun nanofibers. *Artif. Cells Nanomed. Biotechnol.* **2016**, *44*, 847–852. [CrossRef]

51. De Silva, R.T.; Mantilaka, M.M.M.G.P.G.; Goh, K.L.; Ratnayake, S.P.; Amaratunga, G.A.J.; De Silva, K.M.N. Magnesium Oxide Nanoparticles Reinforced Electrospun Alginate-Based Nanofibrous Scaffolds with Improved Physical Properties. *Int. J. Biomater.* **2017**, *2017*, 1391298. [CrossRef]

52. Yang, G.; Xiao, Z.; Long, H.; Ma, K.; Zhang, J.; Ren, X.; Zhang, J. Assessment of the characteristics and biocompatibility of gelatin sponge scaffolds prepared by various crosslinking methods. *Sci. Rep.* **2018**, *8*, 1–13. [CrossRef]

53. Lai, J.-Y. Biocompatibility of chemically cross-linked gelatin hydrogels for ophthalmic use. *J. Mater. Sci. Mater. Med.* **2010**, *21*, 1899–1911. [CrossRef]

54. Blair, T. *Biomedical Textiles for Orthopaedic and Surgical Applications: Fundamentals, Applications and Tissue Engineering*; Woodhead Publishing: Cambridge, UK, 2015.

55. Preda, F.-M.; Alegría, A.; Bocahut, A.; Fillot, L.-A.; Long, D.R.; Sotta, P. Investigation of Water Diffusion Mechanisms in Relation to Polymer Relaxations in Polyamides. *Macromolecules* **2015**, *48*, 5730–5741. [CrossRef]

56. Koudehi, M.F.; Zibaseresht, R. Synthesis of molecularly imprinted polymer nanoparticles containing gentamicin drug as wound dressing based polyvinyl alcohol/gelatin nanofiber. *Mater. Technol.* **2020**, *35*, 21–30. [CrossRef]

57. Aadil, K.R.; Nathani, A.; Sharma, C.S.; Lenka, N.; Gupta, P. Fabrication of biocompatible alginate-poly(vinyl alcohol) nanofibers scaffolds for tissue engineering applications. *Mater. Technol.* **2018**, *33*, 507–512. [CrossRef]

58. Pakdemirli, A.; Toksöz, F.; Karadağ, A.; Misirlioğlu, H.K.; Başbinar, Y.; Ellidokuz, H.; Açikgöz, O. Role of mesenchymal stem cell-derived soluble factors and folic acid in wound healing. *Turk. J. Med. Sci.* **2019**, *49*, 914–921. [CrossRef]

59. Duthie, S.J.; Hawdon, A. DNA instability (strand breakage, uracil misincorporation, and defective repair) is increased by folic acid depletion in human lymphocytes in vitro. *FASEB J.* **1998**, *12*, 1491–1497. [CrossRef]

60. Zhang, C.; Shen, L. Folic acid in combination with adult neural stem cells for the treatment of spinal cord injury in rats. *Int. J. Clin. Exp. Med.* **2015**, *8*, 10471–10480.

61. Datta, P.; Ghosh, P.; Ghosh, K.; Maity, P.; Samanta, S.K.; Ghosh, S.K.; Mohapatra, P.K.D.; Chatterjee, J.; Dhara, S. In vitro ALP and osteocalcin gene expression analysis and in vivo biocompatibility of N-methylene phosphonic chitosan nanofibers for bone regeneration. *J. Biomed. Nanotechnol.* **2013**, *9*, 870–879. [CrossRef]

62. Kamarul, T.; Krishnamurithy, G.; Salih, N.D.; Ibrahim, N.S.; Raghavendran, H.R.B.; Suhaeb, A.R.; Choon, D.S.K. Biocompatibility and Toxicity of Poly(vinyl alcohol)/N,O-Carboxymethyl Chitosan Scaffold. *Sci. World J.* **2014**, *2014*, 905103. [CrossRef]
63. Khalaji, S.; Ebrahimi, N.G.; Hosseinkhani, H. Enhancement of biocompatibility of PVA/HTCC blend polymer with collagen for skin care application. *Int. J. Polym. Mater. Polym. Biomater.* **2021**, *70*, 459–468. [CrossRef]
64. Ekaputra, A.K.; Prestwich, G.D.; Cool, S.M.; Hutmacher, D.W. The three-dimensional vascularization of growth factor-releasing hybrid scaffold of poly (ε-caprolactone)/collagen fibers and hyaluronic acid hydrogel. *Biomaterials* **2011**, *32*, 8108–8117. [CrossRef]
65. Kalalinia, F.; Taherzadeh, Z.; Jirofti, N.; Amiri, N.; Foroghinia, N.; Beheshti, M.; Bazzaz, B.S.F.; Hashemi, M.; Shahroodi, A.; Pishavar, E.; et al. Evaluation of wound healing efficiency of vancomycin-loaded electrospun chitosan/poly ethylene oxide nanofibers in full thickness wound model of rat. *Int. J. Biol. Macromol.* **2021**, *177*, 100–110. [CrossRef] [PubMed]
66. Cetmi, S.D.; Renkler, N.Z.; Kose, A.; Celik, C.; Oncel, S.S. Preparation of electrospun polycaprolactone nanofiber mats loaded with microalgal extracts. *Eng. Life Sci.* **2019**, *19*, 691–699. [CrossRef] [PubMed]
67. Nekounam, H.; Allahyari, Z.; Gholizadeh, S.; Mirzaei, E.; Shokrgozar, M.A.; Faridi-Majidi, R. Simple and robust fabrication and characterization of conductive carbonized nanofibers loaded with gold nanoparticles for bone tissue engineering applications. *Mater. Sci. Eng. C* **2020**, *117*, 111226. [CrossRef] [PubMed]

Article

Preparation of Nanofibers Mats Derived from Task-Specific Polymeric Ionic Liquid for Sensing and Catalytic Applications

David Valverde [1,2], Iván Muñoz [1], Eduardo García-Verdugo [1,*], Belen Altava [1] and Santiago V. Luis [1]

1 Departamento de Química Inorgánica y Orgánica, Universidad Jaume I, Avda. Sos Baynat, s/n, 12071 Castelló de la Plana, Spain; dvalverdeb@uned.ac.cr (D.V.); munozi@uji.es (I.M.); altava@uji.es (B.A.); luiss@uji.es (S.V.L.)
2 Vicerrectoría de Investigación, Universidad Estatal a Distancia, Heredia 40205, Costa Rica
* Correspondence: cepeda@uji.es

Abstract: Nanofibers mats derived from the task-specific functionalized polymeric ionic liquids based on homocysteine thiolactone are obtained by electrospinning them as blends with polyvinylpyrrolidone. The presence of this functional moiety allowed the post-functionalization of these mats through the aminolysis of the thiolactone ring in the presence of an amine by a thiol–alkene "click" reaction. Under controlled experimental conditions the modification can be performed introducing different functionalization and crosslinking of the electrospun fibers, while maintaining the nanostructure obtained by the electrospinning. Initial studies suggest that the nanofibers based on these functionalized polymeric ionic liquids can be used in both sensing and catalytic applications.

Keywords: poly(homocysteine thiolactone); nanofibers; electrospinning; sensing; catalysis

Citation: Valverde, D.; Muñoz, I.; García-Verdugo, E.; Altava, B.; Luis, S.V. Preparation of Nanofibers Mats Derived from Task-Specific Polymeric Ionic Liquid for Sensing and Catalytic Applications. *Polymers* **2021**, *13*, 3110. https://doi.org/10.3390/polym13183110

Academic Editor: Ick-Soo Kim

Received: 27 July 2021
Accepted: 8 September 2021
Published: 15 September 2021

Publisher's Note: MDPI stays neutral with regard to jurisdictional claims in published maps and institutional affiliations.

1. Introduction

Electrospinning techniques combine simplicity, versatility, and low cost with superior capabilities to elaborate scalable ordered and complex nanofiber (NFs) assemblies [1]. NFs mats present advantages over other nanostructured materials due to their high porosity, high surface-area-to-volume ratio, interconnectivity, and ease preparation [2]. These have demonstrated their potential application in different fields such as filter media [3], oil/water separation [4], energy [5], drug delivery [6], sensors [7,8], food packaging [9], tissue engineering [10], catalysis [11], etc.

Polymeric Ionic Liquids (PILs) merge the unique properties of the Ionic Liquids (ILs) with those of advanced materials [12,13]. These macromolecules formed by ionic liquid units connected through a polymeric backbone present an enhanced mechanical stability, processability, flexibility and durability, and improved dimensional control over their structure [14]. PILs with different architectures and properties can be designed according to the desired specific application [15,16]. The combination of the unique properties of the PILs and the use of the electrospinning can lead to nanofibers mats for different applications. However, there are only few examples of the preparation of NFs derived from PILs [17]. In this regard, the molecular structural diversity introduce for PIL has been exploited to modify surface wetting properties of the mats for oil/water separations [18].

PILs nanocomposite fibrous membrane containing metal nanoparticles of tungsten oxide fabricated by electrospinning has been used for the preparation of functional electrolyte for quasi solid-state dye sensitized solar cell (QSS-DSSC) [19]. The power conversion efficiency and stability were improved using electrospun PILs nanofiber composite when compared to the cell with the liquid electrolyte. In a similar way, PILs obtained by chemical modification of poly(vinylidene fluoride-co-hexafluoropropylene) with different molar ratios of 1-butylimidazolium iodide has been explored for the preparation QSS-DSSC [20]. The membrane obtained by electrospinning showed good efficiency with a high η of 9.26%, and excellent long-term stability (up to 97% of its initial η over 1500 h). PILs with well-defined chemical structure were also applied as precursors for nitrogen-doped carbon NFs mats via the electrospinning technique [21], The electrical conductivity of the mat was ca. 200 S cm^{-1}, which is fairly satisfactory compared to carbons prepared from many other organic polymers and materials at the same temperature.

Here, we report on the use of functionalized PILs with amide-thiolactone moieties as a unique platform for the development of NFs mats by electrospinning (Figure 1). The presence of the thiolactone unit allows the orthogonal post-functionalization of these NFs mats through its aminolysis in the presence of an amine (first level of functionalization) followed by the so-called thiol–alkene "click" reaction of the generated –SH groups in the presence of an alkene (second level of functionalization).

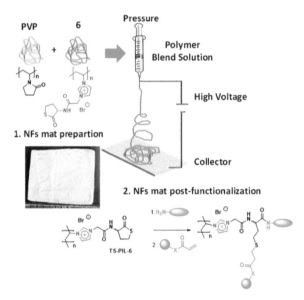

Figure 1. Preparation and post-functionalization of NFs mats derived from TS-PIL **6**.

The high flexibility of this post-modification protocol allows one to obtain a large variety of TS-PILs through the synthetic manipulation of a reduced number of common simple intermediates, easily available even in large scale. The methodology here reported not only allows introducing orthogonal functionalities but also provides a simple strategy for crosslinking the NFs without losing their NFs morphology. The simple preparation and post-medication procedures here developed allows exploiting these NFs mats as advanced materials for sensing and catalytic applications.

2. Materials and Methods

2.1. Materials

All the reagents and solvents used were commercially available. K$_2$CO$_3$ (Scharlau extra pure, Barcelona, Spain), *dl*-homocysteine thiolactone hydrochloride (Aldrich 99%), bromoacetyl bromide (Aldrich 98%, St. Louis, MO, USA), citric acid (Aldrich 99%), anhydrous MgSO$_4$ (Scharlau extra pure), 4,4'-azobis(2-methylpropionitrile) (Aldrich 98%), poly(4-vinylpyridine) (Aldrich, average Mw~160,000 g/mol), and were used as received. The 1-vinylimidazole (Aldrich 99%) was purified by vacuum distillation. All solvents used were of analytic degree.

2.2. General Characterization Protocols

ATR-FTIR spectra were obtained using a spectrometer (JASCO FT/IR-6200) equipped with an ATR (MIRacle single-reflection ATR diamond/ZnSe) accessory at 4 cm^{-1} resolution (4000–600 cm^{-1} spectral range). ^{1}H-NMR experiments were carried out using a Varian INOVA 500 (^{1}H-NMR, 500 MHz, ^{13}C, 125 MHz) and Bruker, Billerica, MA, USA (^{1}H-NMR, 400 MHz) spectrometer.

2.3. Synthetic Protocols

2.3.1. Synthesis of Thiolactone Derivative (3)

dl-homocysteine thiolactone hydrochloride (9.2 g, 60 mmol) was dissolved in milli-Q H$_2$O (20 mL) in a round-bottom flask equipped with a stir bar an ice-cooled (0 °C), then K$_2$CO$_3$ (25,7 g, 186 mmol) was slowly added. The resulting mixture was stirred for 5 min and CH$_2$Cl$_2$ (200 mL) was added. After this, bromoacetyl bromide (10.5 mL, 120 mmol), dissolved in CH$_2$Cl$_2$ (40 mL), was added dropwise during 1 h. After the addition, the reaction mixture was allowed to react for 30 min at 0 °C and then 1 h when the reaction reached rt. A 5% solution of citric acid (2 × 40 mL) was added afterwards to the reaction mixture. The organic phase was decanted, washed with milli-Q$^®$ H$_2$O (2 × 40 mL), dried over anhydrous MgSO$_4$, filtered, and concentrated providing a white solid used without further purification. Yield: 70%. ^{1}H NMR (CD$_3$CN, 400 MHz): δ (ppm): 7.03 (s, 1H), 4.65–4.48 (m, 1H), 3.85 (s, 2H), 3.45–3.21 (m, 2H), 2.58 (dddd, 1H), 2.24–2.06 (m, 1H). ^{13}C NMR (CD$_3$CN, 101 MHz): δ (ppm): 205.7, 167.4, 59.9, 31.1, 29.4, 27.9.

2.3.2. Synthesis of Poly(1-vinylimidazole) (5)

Compound **5** was obtained as shown in Scheme 1 according to methods already reported in the literature. Characterization data were consistent with published values [21].

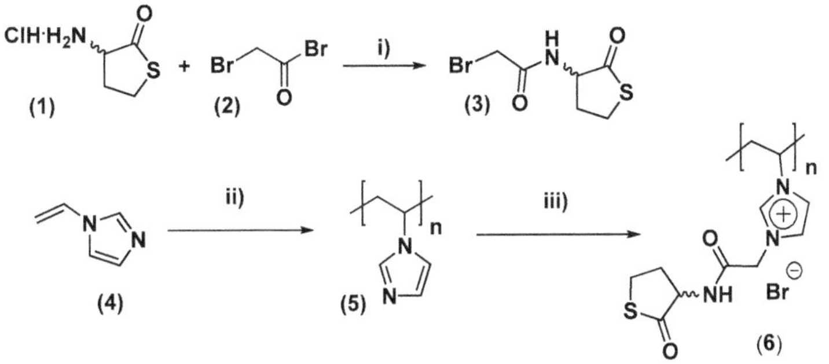

Scheme 1. Synthesis of PIL **6**. (i) K$_2$CO$_3$, H$_2$O/CH$_2$Cl$_2$ 0 °C, then rt. (ii) AIBN, 100 °C, 24 h DMF. (iii) DMF, 80 °C, 24 h.

2.3.3. Quaternization of Poly(1-vinylimidazole) (5) to Produce (6)

Poly(-1-vivylimidazole) (**5**) (5.9 g, 5.88 mmol) was dissolved in dry DMF (15 mL). The solution was heated to 85 °C, and **3** (2,1 g, 8.82 mmol) dissolved in dry DMF (15 mL) was added dropwise. The reaction mixture was bubbled by N$_2$ and was allowed to react for 24 h, to 85 °C with stirring keeping N$_2$ atmosphere. After cooling down to room temperature the polymer was precipitated in to 1 L of diethyl ether and the mixture was left stirring for 30 min. Finally, the diethyl ether was decanted and the solid was washed with diethyl ether (4 × 20 mL) and dried under high vacuum. The solid was purified by dialysis in water and lyophilized, obtaining a brown solid. Yield: 40%. ^{1}H NMR (400 MHz, D$_2$O) δ (ppm): 7.58 (d, 1H), 5.11 (s, 1H), 4.36 (s, 1H), 3.51 (d, 1H), 2.55 (d, 2H). ^{13}C NMR (101 MHz, D$_2$O) δ (ppm): 209.93, 166.27, 137.02, 136.21, 126.40, 119.17, 59.70, 51.16, 39.50, 29.70, 27.84.

2.4. Preparation of the Electrospun Fibers

The electrospinner used was a Fluinateck LE 100.V1 of BioInicia, Paterna, Spain. The polymer solution was loaded into a plastic syringe equipped with a stainless steel needle tip of 0.9 mm inner diameter. The tip-collector working was fixed at x: 15.5 cm, y: 15.5 cm, z: 20.0 cm. The applied voltage was kept between +19.5 and -14.7 kV and the flow rate was 1500 µL min^{-1}. The electrospun fibers were collected on aluminum foil that covered a rotating collector (100 rev min^{-1}). All experiments were carried out at room temperature and relative humidity between 39 and 41%. The obtained membranes were vacuum dried overnight in an oven at 50 °C.

3. Results and Discussion

3.1. Synthesis and Characterization of a Task Specific PIL Containing Thiolactone Fragments

The desired functionalized PIL containing amino-homocysteine thiolactone units was obtained as depicted in Scheme 1. TS-PILs **6** was synthesized by alkylation of poly(1-vinylimidazole) (**5**) with the 2-bromoacetamide of the amino-homocysteine thiolactone **3** in DMF. Poly(1-vinylimidazole) (**5**) was prepared following the methodology reported by Yuan and co-workers [21]. The alkylating agent could be synthetized in multi-gram scale by reaction of dl-homocysteine thiolactone hydrochloride (**1**) with 2-bromoacetyl bromide (**2**). This methodology allows for the synthesis of functionalized PILs with high molecular weight, which are difficult to access by direct polymerization of the corresponding functionalized IL-monomers. Indeed, it should be mentioned that all attempts to obtain the polymer **6** by direct polymerization of the IL-monomer resulting from the alkylation of vinyl imidazole with **3** were unsuccessful.

The chemical modification of the polymer **5** was confirmed by ATR-FTIR and NMR analysis. Figure 2a shows the C=C/C=N stretching bands at 1484 cm^{-1} and 1440 cm^{-1} assignable to the presence of the imidazole rings in polymer **5**. The disappearance of these bands suggests the complete conversion of these imidazole rings into the corresponding imidazolium salts (Figure 2b).

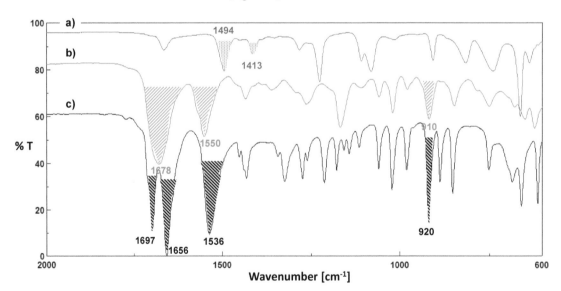

Figure 2. Synthesis of the polymer **6** monitored by ATR-FTIR. (**a**) Polymer **5**, (**b**) Polymer TS-PIL **6** and (**c**) alkylating agent containing amino-thiolactone units (**3**).

The IR-spectra of the polymer **6** also showed the appearance of a broad band at 1650–1720 cm^{-1} associated to the C=O stretching of both thiolactone and amide carbonyl groups, appearing as two different bands for **3** (1697 cm^{-1} and 1656 cm^{-1}, respectively). The polymer **6** also showed an intense peak at 910 cm^{-1} assignable to the C-S stretching of the thiolactone ring.

The ^{1}H-NMR of polymers **5** and **6** confirmed a quantitative quaternization of the imidazole rings (Figure 3). The proton signals of the imidazole ring in **5** (6.4–7.5 ppm) disappeared after the alkylation and shifted to 7.5–9.5 ppm in good agreement with the position expected for imidazolium signals [21]. A series of new signals corresponding to the protons of the -CH$_2$- and –CH groups in the thiolactone units appeared at 4.7–5.4 ppm. The integrals for the thiolactone proton signals at 4.7–5.4 ppm and the imidazolium and amide proton signals (7.5–9.5 ppm) displayed the expected ratio of 4.0/3.0. The presence of the peak at 209 ppm associated with C=O of thiolactone and the peak at 166 ppm assignable to the C=O of the amide on the ^{13}C-NMR of polymer **6** confirmed successful alkylation of the polymeric imidazole (Figure S14).

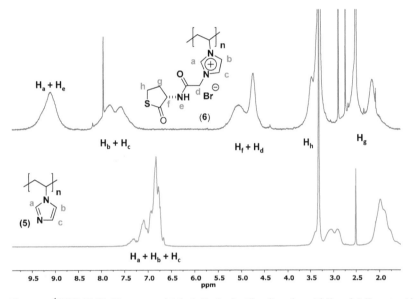

Figure 3. ^{1}H-NMR (D$_2$O) spectra of **5** (poly(1-vinylimidazolium bromide)) and **6** (3-acetamide-*dl*-homocysteine thiolactone-1-vinylimidazolium bromide).

3.2. Synthesis of NFs Mats by Electrospinning

The preparation of NFs mats was evaluated using different electrospinning parameters, which could be adjusted during the process by collecting small samples in-situ and analyzing their morphology by optical microscope. The Table S1 summarizes the conditions evaluated to achieve NFs formation by electrospinning. Negative voltages were applied to the collector to direct the electrospinning jet towards the collector. Once stabilized under the efficient electrospinning conditions, NFs mats were collected over aluminum foil as membranes of ca. 20 × 15 cm^2. The mats were dried under vacuum at 50 °C and their morphology analyzed by scanning electron microscopy (SEM).

Initially, electrospinning of a 65% *w/v* solution of polymer **6** in DMF was assayed. The material collected under these conditions was basically composed of beads mixed with some fibers (Figure S1). At the view of these initial results, a 50% weight/volume polymer blend solution formed by polyvinylpyrrolidone (PVP, Mw = 1,300,000 g/mol) and the polymer **6** in a 1:1 weight ratio in DMF was assayed. It has been reported that PVP-PILs blends can enable the production of NFs by electrospinning [18]. Using this solution and the conditions summarized in Table S1 the collection of NFs mats (NFm-**6**) was possible. Figure 4 shows some representative images of the morphologies observed for these electrospun mats. The PVP/**6** blend afforded a dense entanglement of bead-free fibers in most cases displaying a cylindrical form and with an average diameter of 474 nm (Figure 4a,c). The presence of some micro-ribbons instead of cylindrical fibers was also detected (Figure 4b). The width of these micro-ribbons was in the micrometer range (ca. 2.4 µm). The presence of these ribbons can be attributed to the association of some of the fibers [1]. The mat shows many interconnected NFs crossing in different direction defining a clear open porosity. The section of the NFs mat showed a thickness of ca. of 92 µm (Figure 4d).

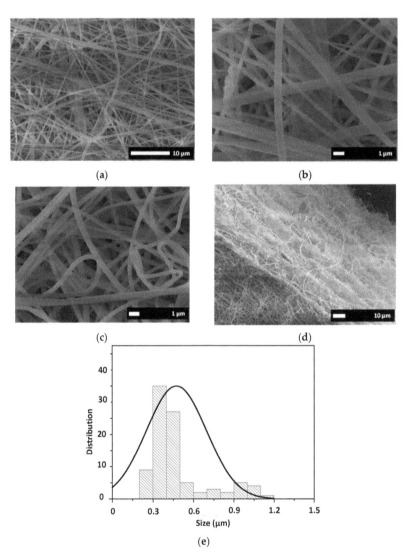

Figure 4. SEM images for the electrospun mat (NFm-**6**) formed from polymer blends PVP/TS-PIL **6** at different magnification. Surface view (**a**–**c**) and section view (**d**). Scale bar represent 10 µm (**a,d**) and 1 µm (**b,c**). (**e**) NFs size distribution.

The obtained NFm-**6** was also analyzed by ATR-FTIR showing the characteristics bands corresponding to both PVP and **6**. A complex broad band at 1750–1600 cm^{-1} was observed for the NFm-**6**, resulting from the overlapping of the individual bands corresponding to the carbonyl groups of the thiolactone and the amide fragments present in **6** and the amide group from PVP. The mat also showed the peak assignable to the amide II of **6**, which appeared red shifted (1553 cm^{-1} vs. 1550 cm^{-1} for **6** alone). The thiolactone C-S band at 920 cm^{-1} also confirmed the presence of **6** in NFm-**6**. The skeletal symmetric stretching of the imidazolium ring along with components assigned to CH$_2$(N) stretching appeared in the mat at 1169 cm^{-1}, while peaks at 1157 cm^{-1} and 1167 cm^{-1} were found for the corresponding original PILs. The peak for the -CH- wagging ν(C–N) of PVP for NFm-**6** mat also showed a shift to 1289 cm^{-1} from the value of 1287 cm^{-1} for pure PVP. All these red shifts suggest an interaction between PVP and **6** leading to an intimate blending through electrospinning. This interaction justifies the morphology of the NFs observed by SEM, where no signals of phase separation were appreciated [22].

3.3. Post-Functionalization of the NFs Mat

Once demonstrated the suitability of polymer **6** blended with PVP for the preparation of NFs mats, their solubility and the effect of different solvents was evaluated. The results obtained are summarized on Table S2. The mat was soluble in polar solvents (water and methanol) while being insoluble in apolar media such as toluene, hexane or diethyl ether. In these solvents, the material maintained their shape (1 × 1 cm mat pieces) once dried. However, when dichloromethane was tested the size of the mat was reduced from 1 × 1 cm to 0.3 × 0.2 cm also increasing its rigidity. Thus, apolar solvents such as toluene and diethyl ether were suitable solvent media for the modification and washing of these mats. Thus, the post-functionalization of **6** with different amines and acrylates and acrylamides was assayed as highlighted in Figure 1. In a first experiment, the modification of NFm-**6** was performed in the presence of an amine by suspending the mats in the corresponding liquid amine at 60 °C for 20 h. The film was then rinsed with toluene and the excess solvent evaporated at 50 °C until constant weight.

Mat **7** was obtained by modification of NFm-**6** with 1,3-diaminopropane (Figure 5). The modification was confirmed by ATR-FTIR (Figure S3a) showing the disappearance of the peak at 920 cm^{-1} assignable to the C-S stretching of the thiolactone ring together with the presence of a strong band at 1557 cm^{-1} (amide II) assignable to the formation of new amide groups. The SEM pictures (Figure 6a) for mat **7** suggest that the post-modification of the membrane induced important changes on the morphology. The modified mat showed a complete porosity loss (Figure 6a vs. Figure 4). The di-topic nature of the amine can yield not only the aminolysis of the thiolactone but also the crosslinking of the polymeric NFs. An additional crosslinking of the NFs via S-S bridges is also possible. Thus, this double crosslinking mechanism by diamide and disulfide bridges led to strong NFs fusion inducing a complete loss of the initial structure. Similarly, the ATR-FTIR of the mat **8** confirmed the post-functionalization with butylamine (Figure 6b). SEM pictures also showed the fusion of NFs and the loss of the porosity. In this case, the change on morphology can be only due to the dynamic crosslinking and fusion of the NFs via S-S bridges (Figure 6b). It is noteworthy that the mats modified under these conditions changed their solubility properties, becoming insoluble in DMF.

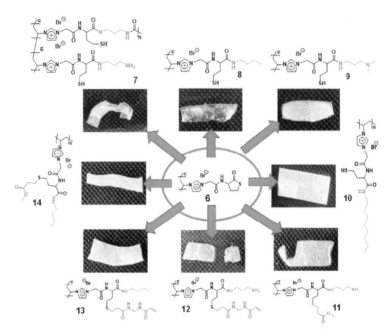

Figure 5. Post-modification of NFm-**6** using different reagents (amines and diamines). Pictures show the appearance of the mat after modification.

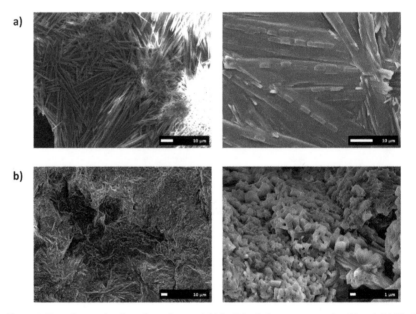

Figure 6. Sem pictures for the polymeric mats (**a**) Mat **7** (scale bars correspond to 10 μm) (**b**) Mat **8** (scale bars correspond to left: 10 μm, right: 1 μm).

ATR-FTIR spectra also demonstrated the successful modification of NFm-**6** with octylamine and 3-(dimethylamine)-1-propylamine affording mats **9** and **10** (Figures S5a and S6a). However, in this case, after the post-functionalization, the open porosity defined by the NFs in NFm-**6** was preserved (Figure 7a,b vs. Figure 4). The modification with 3-(dimethylamine)-1-propylamine produced an increase of the average diameter of the NFs from 474 nm (NFm-**6**) to 644 nm (Figure S5c). Similar results were also found for the mat modified with octylamine, although in this case with a slightly lower NFs size increase (545 nm, Figure S6c). Mats **9** and **10** were insoluble in DMF confirming the crosslinking of the NFs via S-S bridging of the thiols generated in-situ by the aminolysis of the thiolactone moieties. In comparison with the mat **8**, the larger hindrance introduce by the amine, can preclude, at some extent, the crosslinking via S-S bridge decreasing the deformation of the NFs and the loss of the open pore nanostructure of the mat.

Figure 7. Sem pictures for the polymeric mats (**a**) Mat **9** (scale bars correspond to left: 10 μm, right: 1 μm) (**b**) Mat **10** (scale bars correspond to 1 μm).

NFm-**6** could also be modified sequentially with 1,3-diaminopropane and methacrylate (mat **11**), which can react with the thiols obtained in-situ by aminolysis of the thiolactone group and avoiding additional crosslinking via S-S bridges and leading to two different levels of functionalization. The disappearance of the band at 920 cm^{-1} and the presence, among others, of a stronger amide II band at 1556 cm^{-1} confirmed the aminolysis, while the strong peak corresponding to the C=O of the ester at 1730 cm^{-1} suggested the success functionalization of the thiol (Figure S7a). SEM pictures of the modified mat showed that the NFs were fused and entangled together due to crosslinking induced by the diamine (Figure 8). Indeed, the NFs average size increased from 474 nm for the unmodified NFm-**6** to 922 nm. However, in this case, the addition of the acrylate avoided the crosslinking via S-S bridges allowing tus o keep the nanostructure of open pores of the mat (Figure 8 vs. Figure 4), although the porosity was greatly reduced in comparison with the unmodified NFm-**6**.

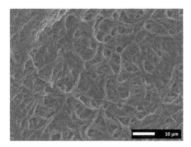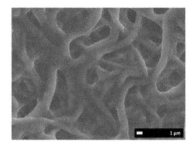

Figure 8. SEM pictures for the polymeric mat **11** (scale bars correspond to 10 μm (**left**), 1 μm (**right**)).

In the search of the additional methodologies allowing the simultaneous modification and crosslinking of the NFs while maintaining the nanostructure and open porosity the second level modification with *N*,*N*′-methylenebis(acrylamide) was evaluated. The bis-acrylamide, in principle, can react through a Michael reaction with the generated thiol groups providing an additional possibility for crosslinking. Two different amines, 1,3-diaminopropane and butyl amine, were assayed for the first level of functionalization. The first combination can lead to crosslinking both through the aminolysis and through thiol modification, while in the case of butyl amine the crosslinking would be only due to the thiolene click reactions. The modification was followed by ATR-FTIR and afforded the modified mats **12** and **13** (Figures S8a and S9a). The mat modified with 1,3-diaminopropane and bisacrylamide showed an almost completely loss of the porosity (Figure S8b). However, thick fused cylindrical NFs were still observable. This change in the morphology can be attributed to the extensive crosslinking provided by the bifunctional modifiers. When the modification was performed with butyl amine, instead of 1,3-diaminopropane, a complete loss of the NFs morphology was appreciated (Figure S9b). This reveals that S-S bridging represents a more efficient mechanism for NFs fusion with the corresponding loss of the original nanostructure.

Finally, the modification of NFm-**6** was performed with butyl amine and methyl methacrylate to minimize the crosslinking of the fibers. The modification was successfully achieved as the ATR-FTIR showed again the band for the ester group at 1738 cm^{-1} and a strong amide II band at 1557 cm^{-1} along with the total disappearance of the C-S peak of the thiolactone group (Figure S10a). The mat after the modification showed entangled cylindrical NFs with an average size of 710 nm, thus preserving the essential nanostructure but achieving a reduction in the porosity in comparison with the morphology observed for unmodified NFm-**6**.

In summary, the presence of the thiolactone unit allows the chemical post-modification of the NFs, although the resulting nanostructure of mats obtained was highly dependent on the nature of the modifiers.

3.4. Application of the NFs Mat to Sensing

In the search of possible applications for the materials here prepared, a series of proof-of-concept studies were carried out. Materials composed by NFs mats have found promising applications as sensing devices [23], for instance, for a fast and low-cost detection of volatile amines [24]. Natural volatile amines, which are often toxic, can be present in our daily life (i.e., chemical manufacturing, agriculture and farming, release by rotten food or exhalation under certain medical conditions or diseases, etc.) [25,26]. In the same way, ILs and related materials have shown to be of interest in optical sensing technologies [27,28]. Different sensing principles have been reported for volatile amines detection. Among them, the use of pH colorimetric sensors, which change their color upon contact to the volatile amine, represents a facile, cost-effective, and non-power operated material for gas amine detection [29].

For this purpose, NFm-**6** films were impregnated with a solution of bromothymol blue (BTB) in toluene (2 mg in 10 mL). The films were dried under vacuum until constant weight showing a yellow color due to the presence of the BTB in its acid form (Figure 9). To evaluate the ability of these mats as volatile amine sensor, they were exposed to the vapors of different amines (NH_3, methylamine, butylamine, iso-propylamine and piperidine). In all the cases, the NFs mats experienced, independently of the amine nature, a color change from yellow to blue in less than a few seconds, easily observable by the naked eye (see Video S1). The change agrees with those for aqueous solutions of BDB thar are yellow at pH < 7, green under neutral pH and blue at pH > 7. The color change was reversible, and the mats could be used in consecutives cycles of air-amine exposure (see Video S1). It is also noteworthy that the films also responded to the presence of acid gases (HCl) experimenting film shrinking and a color change to violet. The change in size can be explained through an increase in the ionic interactions in the fibers.

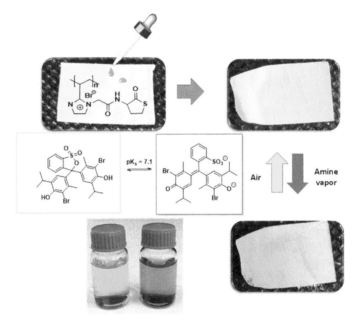

Figure 9. Naked-eye detection of a volatile amine (NH_3) using NFm-**6** doped with bromothymol blue (BTB).

The colorimetric behavior of NFm-**6** was compared, under the same conditions, with that of films prepared by casting from and identical polymeric blend. The sensor obtained by electrospinning, when exposed to the volatile amine, showed a faster and more intense color change than the mat prepared by casting (Figure S11 and Videos S1–S3). Thus, NFm-**6** is able to break the so-called "trade-off" rule in sensing harnessing high sensitivity and fast response simultaneously in comparison with the membrane obtained by casting. The good response relies on the open and well-defined porous structure defined by the NFs (Figure 3) allowing a fast diffusion of the target analyte into the membrane and enhancing its interaction with the sensing unit.

These initial results demonstrated that NFs mats obtained from TS-PILs **6** are a suitable platform for sensing application as: (i) the TS-PIL units do not react or interfere with the sensing dye, (ii) the polymer provides a uniform background not interfering in the color change, (iii) the mats present good surface properties, allowing the colorimetric sensor to spread uniformly, which is reflected in the uniform mat color, and (iv) the NFs structure obtained by electrospinning render a microstructure with high surface area enhancing the diffusion of the analyte to the sensing unit and leading to a fast response.

3.5. Application of the Fiber-Mats for Catalysis

Electrospun NFs presenting catalytic units have demonstrated great potential in a variety of catalytic application [30]. The large surface area and optimal surface chemistry of electrospun nanofibers greatly enhance the catalyst–support interaction and can improve their activity, selectivity, stability, and reusability. Although NFs mats obtained by electrospinning offers very attractive properties for the development of catalytic systems, there is not a single application, as far as we know, of electrospun NFs mats based on PILs and used in catalysis.

Advanced materials based on PILs have been used as efficient systems for the capture, activation and conversion of CO_2 [31]. For instance, different polymers containing IL-like moieties have been reported as efficient catalysts for the cycloaddition reaction of CO_2 with various epoxides to generate cyclic carbonates, which is one of the most promising and efficient approaches for CO_2 fixation [32]. In this context, the modified mat **14** was studied for the catalytic reaction of styrene oxide with CO_2. In general, this process requires harsh experimental conditions (temperature > 130 °C and pressure > 10 bars) to lead to the corresponding carbonates. When the mat **14** was tested as a heterogeneous catalyst for this reaction (Scheme 2) 44% of conversion of the epoxide was observed with full selectivity to the corresponding carbonate at 120 °C and 10 bar after 24 h. This is a remarkable result if we consider the bromide content, as this is expected to be the nucleophilic anion catalyzing the reaction, which was only 0.03% mol Br/substrate. This represents a notable TON of 1244.

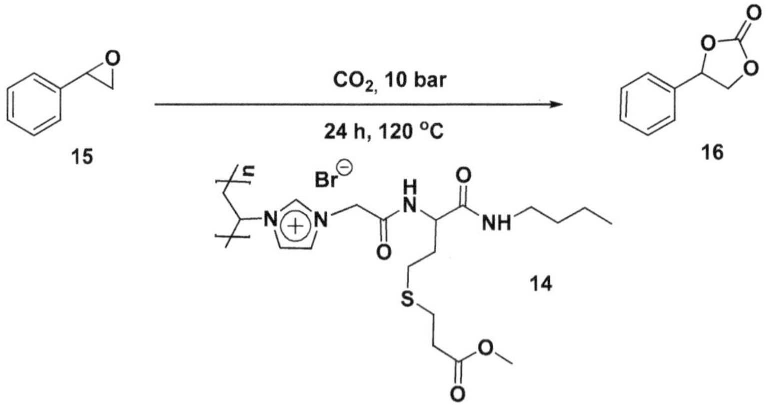

Scheme 2. Cycloaddition reaction of CO_2 to styrene epoxide catalyzed by **14**.

4. Conclusions

We have demonstrated that NFs mats derived from the TS-PILs based on homocysteine thiolactone can be obtained by electrospinning them as blends with PVP. The presence of this functional moiety allowed the post-functionalization of these mats through the aminolysis of the thiolactone ring in the presence of an amine that can be followed by a thiol–alkene "click" reaction. Under controlled experimental conditions the modification can be performed introducing different functionalization and crosslinking of the NFs. The morphology of the modified mats was highly dependent of nature of the modifiers. Different conditions were stabilized for the modification of the mat while maintaining the nanostructure obtained by the electrospinning. Initial studies suggest that the NFs based on these functionalized PILs can be used in both sensing and catalytic applications.

Supplementary Materials: The following are available online at https://drive.google.com/file/d/12 oPPZsJbuGy0o07QhWyr92Iv_spenN_l/view?usp=drive_web. Figure S1: Images of fibers with the formation of beads obtained by electrospinning of polymer 6 (65 % w/v). (a) Optical microscope imagen, 10 x / 0.25 mm. (b) SEM imagen, scale bar represents 10 μm, Figure S3: (a) Comparison of the ATR-FTIR spectra for NFm-6 and 7. (b) Optical images for mat 7 after modification. and structure, Figure S5: (a) Comparison of the ATR-FTIR for the NFs mat 6 and 9. (b) Optical imagine of the mat 9 after modification and structure. (c) NFs size distribution., Figure S6: (a) Comparison of the ATR-FTIR for the NFs mat 6 and 10. (b) Optical imagine of the mat 10 after modification and structure. (c) NFs size distribution, Figure S7: (a) Comparison of the ATR-FTIR for the NFs mat 6 and 11. (b) Optical imagine of the mat 11 after modification and structure. (c) NFs size distribution, Figure S8: (a) Comparison of the ATR-FTIR for the NFs mat 6 and 12. (b) Optical imagine of the mat 12 after modification and structure. (c) SEM of the polymeric mat 12, scale bars correspond to 10μm, Figure S9: (a) Comparison of the ATR-FTIR for the NFs mat 6 and 13. (b) Optical imagine of the mat 13 after modification and structure. (c) SEM of the polymeric mat 13, scale bars correspond to10μm, Figure S10: (a) Comparison of the ATR-FTIR for the NFs mat 6 and 14. (b) Optical imagine of the mat 14 after modification and structure. (c) SEM of the polymeric mat 14, scale bars correspond to 10μm (d) NFs size distribution, Figure S11: Comparison of color changing of the polymer blend PVP/TS-PIL 6 obtained by (a) casting or by (b) electrospinning in presence of the NH_3 vapor, Figure S14: Spectrum of polymer 6. (a) ^{1}H NMR spectra in D_2O and (b) ^{13}C NMR Spectra in D_2O, Table S1: Electrospinning parameters used to fabricate NFs mats., Table S1: Solubility evaluation of the NFs mat obtained from PVP/TS-PIL 6, Video S1: Colorimetric changes of NFm-6 with butylamine, Video S2: Colorimetric changes of NFm-6 with different amines, Video S3: Colorimetric changes of NFm-6 with ammonia.

Author Contributions: Conceptualization, E.G.-V. and S.V.L.; investigation, D.V. and I.M.; writing-original draft preparation, E.G.-V., B.A. and S.V.L.; writing-review and editing, E.G.-V. and D.V.; supervision, E.G.-V. and B.A.; project administration and funding acquisition, E.G.-V., B.A. and S.V.L. All authors have read and agreed to the published version of the manuscript.

Funding: This work was partially supported by UJI-B2019-40 (Pla de Promoció de la Investigació de la Universitat Jaume I) and RTI2018-098233-B-C22 (FEDER/Ministerio de Ciencia e Innovación–Agencia Estatal de Investigación). D.V. thanks UNED (Costa Rica) for the predoctoral fellowship.

Institutional Review Board Statement: Not applicable.

Informed Consent Statement: Not applicable.

Data Availability Statement: Not applicable.

Acknowledgments: The authors are grateful to the SCIC of the Universitat Jaume I for technical support.

Conflicts of Interest: The authors declare no conflict of interest.

References

1. Xue, J.; Wu, T.; Dai, Y.; Xia, Y. Electrospinning and Electrospun Nanofibers: Methods, Materials, and Applications. *Chem. Rev.* **2019**, *119*, 5298–5415. [CrossRef] [PubMed]
2. Ibrahim, H.M.; Klingner, A. A Review on Electrospun Polymeric Nanofibers: Production Parameters and Potential Applications. *Polym. Test.* **2020**, *90*, 106647. [CrossRef]
3. Ma, Q.; Cheng, H.; Fane, A.G.; Wang, R.; Zhang, H. Recent Development of Advanced Materials with Special Wettability for Selective Oil/Water Separation. *Small* **2016**, *12*, 2186–2202. [CrossRef] [PubMed]
4. Ma, W.; Zhang, Q.; Hua, D.; Xiong, R.; Zhao, J.; Rao, W.; Huang, S.; Zhan, X.; Chen, F.; Huang, C. Electrospun Fibers for Oil-Water Separation. *RSC Adv.* **2016**, *6*, 12868–12884. [CrossRef]
5. Cavaliere, S.; Subianto, S.; Savych, I.; Jones, D.J.; Rozière, J. Electrospinning: Designed Architectures for Energy Conversion and Storage Devices. *Energy Environ. Sci.* **2011**, *4*, 4761–4785. [CrossRef]
6. Sebe, I.; Szabó, P.; Kállai-Szabó, B.; Zelkó, R. Incorporating Small Molecules or Biologics into Nanofibers for Optimized Drug Release: A Review. *Int. J. Pharm.* **2015**, *494*, 516–530. [CrossRef] [PubMed]
7. Liu, Y.; Hao, M.; Chen, Z.; Liu, L.; Liu, Y.; Yang, W.; Ramakrishna, S. A Review on Recent Advances in Application of Elec-trospun Nanofiber Materials as Biosensors. *Curr. Opin. Biomed. Eng.* **2020**, *13*, 174–189. [CrossRef]
8. Zampetti, A.E.; Kny, E. *Electrospinning for High Performance Sensors*; Springer: Cham, Switzerland, 2015.
9. Mohammadi, M.A.; Hosseini, S.M.; Yousefi, M. Application of Electrospinning Technique in Development of Intelligent Food Packaging: A Short Review of Recent Trends. *Food Sci. Nutr.* **2020**, *8*, 4656–4665. [CrossRef]

10. Pham, Q.P.; Sharma, U.; Mikos, A.G. Electrospinning of Polymeric Nanofibers for Tissue Engineering Applications: A Review. *Tissue Eng.* **2006**, *12*, 1197–1211. [CrossRef]
11. Huang, Z.-M.; Zhang, Y.-Z.; Kotaki, M.; Ramakrishna, S. A Review on Polymer Nanofibers by Electrospinning and Their Applications in Nanocomposites. *Compos. Sci. Technol.* **2003**, *63*, 2223–2253. [CrossRef]
12. Zhang, S.-Y.; Zhuang, Q.; Zhang, M.; Wang, H.; Gao, Z.; Sun, J.-K.; Yuan, J. Poly(Ionic Liquid) Composites. *Chem. Soc. Rev.* **2020**, *49*, 1726–1755. [CrossRef]
13. Qian, W.; Texter, J.; Yan, F. Frontiers in Poly(Ionic Liquid)s: Syntheses and applications. *Chem. Soc. Rev.* **2017**, *46*, 1124–1159. [CrossRef]
14. Montolio, S.; Altava, B.; García-Verdugo, E.; Luis, S.V. Supported ILs and materials based on ILs for the development of green synthetic processes and procedures. In *Green Synthetic Processes and Procedures*; RSC Publishing: London, UK, 2019; pp. 289–318.
15. Lin, H.; Zhang, S.; Sun, J.-K.; Antonietti, M.; Yuan, J. Poly(Ionic Liquid)s with Engineered Nanopores for Energy and Environmental Applications. *Polymer* **2020**, *202*, 122640. [CrossRef]
16. Minami, H. Preparation and Morphology Control of Poly(Ionic Liquid) Particles. *Langmuir* **2020**, *36*, 8668–8679. [CrossRef] [PubMed]
17. Josef, E.; Guterman, R. Designing Solutions for Electrospinning of Poly(Ionic Liquid)s. *Macromolecules* **2019**, *52*, 5223–5230. [CrossRef]
18. Montolio, S.; Abarca, G.; Porcar, R.; Dupont, J.; Burguete, M.I.; García-Verdugo, E.; Luis, S.V. Hierarchically Structured Polymeric Ionic Liquids and Polyvinylpyrrolidone Mat-Fibers Fabricated by Electrospinning. *J. Mater. Chem. A* **2017**, *5*, 9733–9744. [CrossRef]
19. Thomas, M.; Rajiv, S. Grafted PEO Polymeric Ionic Liquid Nanocomposite Electrospun Membrane for Efficient and Stable Dye Sensitized Solar Cell. *Electrochim. Acta* **2020**, *341*, 136040. [CrossRef]
20. Pang, H.-W.; Yu, H.-F.; Huang, Y.-J.; Li, C.-T.; Ho, K.-C. Electrospun Membranes of Imidazole-Grafted PVDF-HFP Poly-meric Ionic Liquids for Highly Efficient Quasi-Solid-State Dye-Sensitized Solar Cells. *J. Mater. Chem. A* **2018**, *6*, 14215–14223. [CrossRef]
21. Yuan, J.; Márquez, A.G.; Reinacher, J.; Giordano, C.; Janek, J.; Antonietti, M. Nitrogen-Doped Carbon Fibers and Mem-branes by Carbonization of Electrospun Poly(Ionic Liquid)s. *Polym. Chem.* **2011**, *2*, 1654–1657. [CrossRef]
22. Bahadur, I.; Momin, M.I.K.; Koorbanally, N.A.; Sattari, M.; Ebenso, E.E.; Katata-Seru, L.M.; Singh, S.; Ramjugernath, D. Interactions of Polyvinylpyrrolidone with Imidazolium Based Ionic Liquids: Spectroscopic and Density Functional Theory Studies. *J. Mol. Liq.* **2016**, *213*, 13–16. [CrossRef]
23. Schoolaert, E.; Hoogenboom, R.; Clerck, K.D. Colorimetric Nanofibers as Optical Sensors. *Adv. Funct. Mater.* **2017**, *27*, 1702646. [CrossRef]
24. Jornet-Martínez, N.; Moliner-Martínez, Y.; Herráez-Hernández, R.; Molins-Legua, C.; Verdú-Andrés, J.; Campíns-Falcó, P. Designing Solid Optical Sensors for in Situ Passive Discrimination of Volatile Amines Based on a New One-Step Hydrophilic PDMS Preparation. *Sens. Actuators B* **2016**, *223*, 333–342. [CrossRef]
25. Askim, J.R.; Mahmoudi, M.; Suslick, K.S. Optical Sensor Arrays for Chemical Sensing: The Optoelectronic Nose. *Chem. Soc. Rev.* **2013**, *42*, 8649–8682. [CrossRef]
26. Li, Z.; Askim, J.R.; Suslick, K.S. The Optoelectronic Nose: Colorimetric and Fluorometric Sensor Arrays. *Chem. Rev.* **2019**, *119*, 231–292. [CrossRef]
27. Muginova, S.V.; Myasnikova, D.A.; Kazarian, S.G.; Shekhovtsova, T.N. Applications of Ionic Liquids for the Development of Optical Chemical Sensors and Biosensors. *Anal. Sci.* **2017**, *33*, 261–274. [CrossRef] [PubMed]
28. Guterman, R.; Ambrogi, M.; Yuan, J. Harnessing Poly(Ionic Liquid)s for Sensing Applications. *Macromol. Rapid Commun.* **2016**, *37*, 1106–1115. [CrossRef]
29. Ruiz-Capillas, C.; Jiménez-Colmenero, F. Biogenic Amines in Meat and Meat Products. *Crit. Rev. Food Sci. Nutr.* **2005**, *44*, 489–599. [CrossRef] [PubMed]
30. Lu, P.; Murray, S.; Zhu, M. Electrospun Nanofibers for Catalysts. In *Electrospinning: Nanofabrication and Applications*; Ding, B., Wang, X., Yu, J., Eds.; Micro and Nano Technologies; William Andrew Publishing: Norwich, NY, USA, 2019; pp. 695–717.
31. Zhou, X.; Weber, J.; Yuan, J. Poly(Ionic Liquid)s: Platform for CO_2 Capture and Catalysis. *Curr. Opin. Green Sustain. Chem.* **2019**, *16*, 39–46. [CrossRef]
32. Luo, R.; Liu, X.; Chen, M.; Liu, B.; Fang, Y. Recent Advances on Imidazolium-Functionalized Organic Cationic Polymers for CO_2 Adsorption and Simultaneous Conversion into Cyclic Carbonates. *ChemSusChem* **2020**, *13*, 3945–3966. [CrossRef]

MDPI

Article

Selective Adsorption and Separation of Proteins by Ligand-Modified Nanofiber Fabric

Song Liu and Yasuhito Mukai *

Department of Chemical Systems Engineering, Nagoya University, Furo-cho, Chikusa-ku, Nagoya 4648603, Japan;
liusong@morimatsu.cn
* Correspondence: mukai.yasuhito@material.nagoya-u.ac.jp

Abstract: Electrospun polyvinyl alcohol (PVA) nanofiber fabric was modified by Cibacron Blue F3GA (CB) to enhance the affinity of the fabric. Batch experiments were performed to study the nanofiber fabric's bovine hemoglobin (BHb) adsorption capacity at different protein concentrations before and after modification. The maximum BHb adsorption capacity of the modified nanofiber fabric was 686 mg/g, which was much larger than the 58 mg/g of the original fabric. After that, the effect of feed concentration and permeation rate on the dynamic adsorption behaviors for BHb of the nanofiber fabric was investigated. The pH impact on BHb and bovine serum albumin (BSA) adsorption was examined by static adsorption experiments of single protein solutions. The selective separation experiments of the BHb–BSA binary solution were carried out at the optimal pH value, and a high selectivity factor of 5.45 for BHb was achieved. Finally, the reusability of the nanofiber fabric was examined using three adsorption–elution cycle tests. This research demonstrated the potential of the CB-modified PVA nanofiber fabric in protein adsorption and selective separation.

Keywords: nanofiber fabric; protein; affinity adsorption; selective separation

Citation: Liu, S.; Mukai, Y. Selective Adsorption and Separation of Proteins by Ligand-Modified Nanofiber Fabric. *Polymers* **2021**, *13*, 2313. https://doi.org/10.3390/polym13142313

Academic Editor: Francisco Javier Espinach Orús

Received: 21 June 2021
Accepted: 11 July 2021
Published: 14 July 2021

Publisher's Note: MDPI stays neutral with regard to jurisdictional claims in published maps and institutional affiliations.

1. Introduction

Proteins are biological macromolecules that maintain the integrity of the cell structure and function [1]. Proteins are abundant in biology and are essential building blocks of life [2]. All metabolic activities of living organisms, including growth, development and reproduction, are related to the activity and metabolism of proteins. The cells of living organisms consist of different kinds of proteins, each of which has a different structure and function [3–5]. Proteins generally exist in complex mixed forms in tissues and cells, and various types of cells contain thousands of different proteins. As proteins vary in size, charge and water solubility, it is challenging to obtain target proteins from mixtures [6,7]. Protein separation and purification have a wide range of applications in our daily life, such as dairy [8], food industry [9], biomedicine [10], etc. The overall goal of protein isolation and purification is to increase the purity of the product and to maintain the isolated protein's biological activity. It is desirable to achieve high production efficiency and high product quality while reducing the cost of the isolation process.

Precipitation, filtration, dialysis and chromatography techniques have been utilized for industrial separations [11–14]. Filtration is one of the most broadly used processes in protein separation. Filtration separation is based on the variability of the size and shape of protein molecules [15], where the protein solution is driven by a pressure difference to flow through the porous filter media. Protein molecules for which the sizes are smaller than the filter pore size pass through the filter medium, while protein molecules of larger sizes are retained, resulting in the separation and purification of proteins [16,17]. Taking the widely used ultrafiltration membrane as an example, the mass transfer resistance of proteins through the ultrafiltration membrane is large, which leads to high resistance. During the ultrafiltration process, pore blockage and fouling usually lead to a gradual decrease in treatment flux [18]. Furthermore, conventional ultrafiltration is limited to separating solutes with molecular size differences of more than 10 times [19]. The separation selectivity is hardly controllable for protein molecules of similar sizes [20].

The tedious process of biological product recovery dramatically impacts the final product cost because separation and purification account for 70% to 80% of the total biological process cost [21]. The significant growth in the protein market poses a considerable challenge to industrial production, particularly downstream processing, due to its limited production capacities and high cost. It is necessary to develop simple, rapid, scalable and economically feasible methods for protein separation. Polymer nanomaterials technology has brought a profound impact on the fields of artificial intelligence [22], telecommunication [23] and energy [24] and has provided a direction for the solution to the troubles in protein separation. Electrospun nanofibers are characterized by large specific surface ratios, high porosity (over 80%), superior mechanical strength, excellent controllability of the spinning process and a wide variety of raw materials [25–27]. It has attracted great attention from researchers in the field of adsorption and separation [28–30].

Up until now, a series of electrospun nanofibers has been successfully prepared for the adsorption and separation of protein [31–33]. Much smaller fiber diameters and increased surface areas are accessible through the electrospinning technique compared with currently available textile fibers [34]. Additionally, the large surface area provides abundant sites for protein adsorption. Electrospun nanofibers are ideal materials for a protein affinity medium because of their adjustable functionality and open porous structures. After modification by affinity ligands, the resulting nanofiber fabrics can capture target protein molecules based on highly specific binding interactions rather than on molecular weight, size or charge. Duan et al. [35] fabricated a novel metal-chelating affinity nanofiber fabric by methodically modifying the cellulose nanofiber fabric with an intermediate bridging agent of cyanuric chloride, chelating the agent of iminodiacetic acid (IDA) and affinity ligand of Fe^{3+} ions. The prepared Fe^{3+} ions' immobilized nanofiber fabric presented a high lysozyme adsorption capacity of 365 mg/g. Reactive dyes can interact with the active sites of many proteins by mimicking the structure of the substrate or cofactors of these proteins and are excellent ligand candidates for affinity [36]. Ng et al. [37] developed a chitosan and Reactive Orange 4 modified polyacrylonitrile nanofiber fabric. The obtained dye affinity fabric was used to separate the lysozyme from chicken egg whites, and the total recovery ratio of lysozyme was up to 98.6%, with a purification multiple of 56.9 times. In two recent reports from our laboratory [38,39], we enhanced the affinity of the polyvinyl alcohol (PVA) nanofiber fabrics using Cibacron Blue F3GA (CB), which is a dye with specificity and binding abilities to a series of proteins. Then, the static and dynamic adsorption and desorption behaviors of the PVA nanofiber fabrics for bovine serum albumin (BSA) were investigated. The CB-enhanced PVA nanofiber fabric had a BSA adsorption capacity of 769 mg/g. To continue the research on protein separation, we are committed to the separation of proteins with similar sizes by nanofiber fabric. This research is not limited to the adsorption of a single protein and is more informative about the separation process of protein mixtures.

In this research, bovine hemoglobin (BHb) and BSA were used as model proteins for selective separation. Hemoglobin is a carrier of oxygen and aids the transport of carbon dioxide, and BHb has more than 85% amino acid sequence homology with human hemoglobin [40]. Serum albumin maintains the osmotic pressure and pH of the blood and transports a wide variety of endogenous and exogenous compounds, and BSA displays approximately 76% sequence homology with human serum albumin [41]. The molecular weights of BHb and BSA are 64,500 and 67,000, respectively [42,43]. The sizes of BHb and BSA are too close to be effectively separated by conventional filtration operations. First, the PVA nanofiber fabric was modified with CB to improve its affinity. Next, the static and dynamic adsorption performances of the PVA nanofiber fabrics for BHb before and after the CB modification were investigated. Then, the effect of pH on the BHb and BSA adsorption performance was examined. Finally, the selective separation experiments of

BHb and BSA were carried out at the optimal pH value, with the reusability of the PVA nanofiber fabrics also being studied.

2. Materials and Methods

2.1. Materials

Electrospun polyvinyl alcohol (PVA) nanofiber fabrics were obtained from Japan Vilene Company, Ltd., Koga, Japan. The molecular weight of the PVA is 66,000–79,000, and the mass per unit area of the PVA nanofiber fabric is 40 g/m². Cibacron Blue F3GA (CB) was supplied by Polysciences, Inc., Tokyo, Japan. Bovine hemoglobin (BHb) and bovine serum albumin (BSA) were purchased from Sigma-Aldrich Co. LLC., Tokyo, Japan. The other chemicals were supplied by Wako Pure Chemical Industries, Ltd., Osaka, Japan.

2.2. Preparation of CB-Modified PVA Nanofiber Fabric

The CB-modified PVA nanofiber fabrics were prepared according to previously reported methods [38,44,45]. One hundred micrograms of CB was dissolved in 10 mL of water to prepare a dyeing solution. Then, the dyeing solution was heated to 60 °C, and a PVA nanofiber fabric was soaked in the solution for 30 min. After that, 2 g of NaCl was added to stimulate the deposition of the dye on the internal surface of the PVA nanofiber fabric. After maintaining the fabric at 60 °C for 1 h, 0.2 g of Na_2CO_3 sodium carbonate was added to adjust the solution's pH value to accelerate the reaction between the dye and the PVA nanofiber fabric, which took place at 80 °C for 2 h. Finally, the dyed PVA nanofiber fabric was thoroughly washed with deionized water until the absence of CB molecules in the wash was detected by measuring the UV-vis absorbance at 600 nm (the maximum absorption wavelength of CB).

2.3. BHb Adsorption Studies

The BHb adsorption capacities of the original and CB-modified PVA nanofiber fabrics were measured under different initial BHb concentrations at pH = 6.8 (BHb's isoelectric point). The two types of fabrics were immersed into an 8 mL BHb solution with concentrations varying from 1 g/L to 6 g/L. The static adsorption experiments were carried out at 25 °C for 6 h. The BHb solution concentrations were detected by a UV-vis spectrophotometer at 406 nm before and after adsorption. To analyze the adsorption results, the amount of adsorbed protein per unit mass of fabric was determined with the following Equation:

$$q = \frac{(C_0 - C)V}{W} \tag{1}$$

where q is the adsorption capacity (mg/g); C_0 and C are the initial and the final BHb concentrations (mg/L), respectively; V is the volume of the solution (L); and W is the fiber mass (g). The BHb adsorption capacities were plotted against the final concentrations to obtain an adsorption curve.

The dynamic adsorption performance of BHb on the PVA nanofiber fabrics was studied by permeation tests. Figure 1 depicts the experimental setup. The PVA nanofiber fabric with a weight of 19.6 mg was placed in the Millipore filter holder (SX0002500) with an effective diameter of 2.2 cm and a permeation area of 3.8 cm². Then, the inlet of the filter holder was mounted to the outlet of the syringe by a threaded connection. Subsequently, a syringe pump (SRS-2, AS ONE Co., Osaka, Japan) was used to inject the BHb solution into the filter holder. After setting the flow rate, the permeation experiment was started by collecting the permeate from the outlet below the filter holder and by monitoring the concentration of BHb. Detailed studies were conducted at 25 °C, BHb feed concentrations of 0.3–1.2 g/L and permeation rates of 2–10 mL/h. The momentary rejection ratio R_m for protein was determined according to the following Equation:

$$R_m = 1 - \frac{C_p}{C_0} \tag{2}$$

where C_0 (g/L) and C_p (g/L) are the protein concentrations in the feed solution and the permeate, respectively.

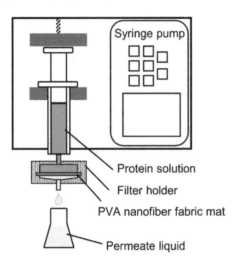

Figure 1. Schematic diagram of the dynamic adsorption apparatus.

At a cumulative permeate volume of V (mL), the rejection ratio R_c for protein of the CB modified PVA nanofiber fabric is given by

$$R_c = \frac{1}{V} \int_0^V R_m dV \qquad (3)$$

2.4. Selective Separation of Binary BHb–BSA Solution

The single-protein adsorption experiments were carried out at 25 °C for 6 h. The single-component concentrations of BHb or BSA solution were 3 g/L. The effect of pH on the adsorption capacity was investigated using different pH values ranging from 4 to 9. The protein concentrations before and after adsorption were measured using a UV-vis spectrometer, referring to the absorption band at 406 nm for BHb and 280 nm for BSA.

Selective experiments were carried out with the permeation apparatus at the optimum pH of 6.8. The initial concentrations of both proteins were 0.6 g/L in the BHb–BSA binary solution. The BHb solution exhibits two characteristic absorbance peaks at 280 nm and 406 nm, while the absorbance of the BSA solution exhibits a maximum around 280 nm and can be negligible around 406 nm. Accordingly, the BHb concentration was determined directly from the absorbance at 406 nm. The BSA concentration was determined at 280 nm by subtracting the contribution of BHb from the concentration detected at 406 nm. The rejection ratio for BHb or BSA is calculated according to Equation (3), and the selectivity factor S of protein is defined by the following equation:

$$S = \frac{R_{c,BHb}}{R_{c,BSA}} \qquad (4)$$

Finally, the reusability tests for BHB–BSA binary protein dynamic adsorption were repeated for three cycles to examine the stability of the CB-modified PVA nanofiber fabrics.

3. Results and Discussion

3.1. Characteristics of the PVA Nanofiber Fabrics

Morphologies of the PVA nanofiber fabrics before and after CB modification were observed by scanning electron microscopy (SEM, S4300, Hitachi High-Technologies Corporation, Tokyo, Japan), operating at 15 kV. Figure 2 shows the SEM images of the PVA nanofiber fabrics. The fiber diameters were analyzed using the ImageJ software. The original PVA nanofiber fabrics were smoother than the CB-modified PVA nanofiber fabrics. After the CB modification, the average fiber diameter increased from the original 224 nm to 238 nm. The CB molecules were covalently immobilized on the fabric surface, which is one of the reasons for the coarsening of the fibers. In addition, the CB modification treatment was carried out in solution, and the hydrophilic PVA nanofiber fabrics probably had a slight swelling.

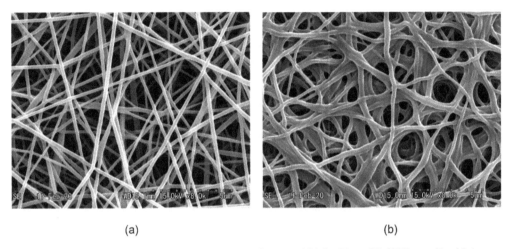

(a)   (b)

Figure 2. SEM images of (**a**) the original PVA nanofiber fabrics and (**b**) the CB-modified PVA nanofiber fabrics.

3.2. Static Adsorption Isotherm of BHb

In order to evaluate the BHb adsorption capacity of the PVA nanofiber fabrics before and after CB modification, the effect of the initial concentration on the adsorption capacity of the nanofiber fabrics was studied, with the results shown in Figure 3.

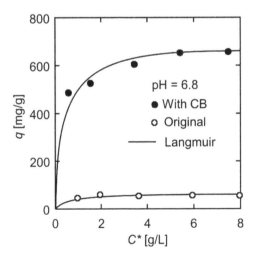

Figure 3. Effect of the BHb initial concentration on adsorption capacities of the original and CB-modified PVA nanofiber fabrics.

The BHb adsorption on the PVA nanofiber fabrics increased with the increase in the BHb concentration. The adsorption capacity of the CB-modified PVA nanofiber fabrics was markedly greater than that of the original fabrics. The adsorption amount of BHb on the original PVA nanofiber fabrics was at a low level in all experimental concentrations, while the adsorption amount of the modified nanofiber fabrics was around 655 mg/g when the BHb concentration exceeded 6.0 g/L. The excellent adsorption capacity of the modified nanofiber fabrics is attributed to the affinity of CB molecules. The interaction between the CB molecule and BHb is a consequence of the combined effects of electrostatic and hydrophobic interactions, and hydrogen bonding. As a monochlorotriazine dye, the CB molecule contains three sulfonic acid groups, and four basic primary and secondary amino groups. The triazine part of CB is used to fix onto the PVA matrix, while the three sulfonic acid groups and the multiple aromatic parts are mainly responsible for the CB–BHb binding [46].

The Langmuir adsorption isotherm is one of the most widely used adsorption isotherms which is used to fit adsorption equilibrium data. It is assumed that adsorption is a monolayer type occurring on the homogeneous surface of the adsorbent [47,48]. The linear form of the Langmuir equation is expressed as follows:

$$\frac{C^*}{q_e} = \frac{C^*}{q_s} + \frac{1}{Kq_s} \tag{5}$$

where q_s (mg/g) is the maximum adsorption capacity, q_e (mg/g) is the equilibrium adsorption capacity, C^* (g/L) is the equilibrium BHb concentration and K (L/g) is the dissociation constant of the system.

By fitting the experimental data to the Langmuir linear equation, the maximum adsorption capacities of BHb on the PVA nanofiber fabrics before and after CB modification were calculated to be 58 mg/g and 686 mg/g, respectively. Zhang et al. [49] prepared magnetic carbon nanotubes with a hierarchical copper silicate nanostructure for BHb adsorption, with the maximum BHb adsorption capacity being 302 mg/g. Wang et al. [50] reported that magnetic mesoporous ytterbium silicate microspheres achieved a BHb adsorption capacity of 304 mg/g. Compared to these materials, the CB-modified PVA nanofiber fabric has a distinct advantage in adsorption capacity.

3.3. BHb Dynamic Adsorption Performance

Dynamic adsorption operation offers the benefits of automatic control, labor saving and easy integration with other continuous processes. Thus, we tested the BHb dynamic adsorption behavior of the CB-modified PVA nanofiber fabrics.

Figure 4a shows the changes in the BHb rejection ratio over the permeation volume from 0 to 10 mL for both the original and CB-modified PVA nanofiber fabrics. The BHb rejection ratio by both nanofiber fabrics showed a downward trend with the increase in permeation volume. Whereas the BHb rejection ratio of the CB PVA nanofiber-modified nanofiber fabric remained at a high level, reducing from 1.0 to 0.62, the BHb rejection ratio of the original nanofiber fabric was meagre, being 0.23 at the beginning and only 0.10 at the end of the experiment. The enhancement in the protein rejection ratio of the CB-modified PVA nanofiber fabric comes from the affinity effect of CB molecules.

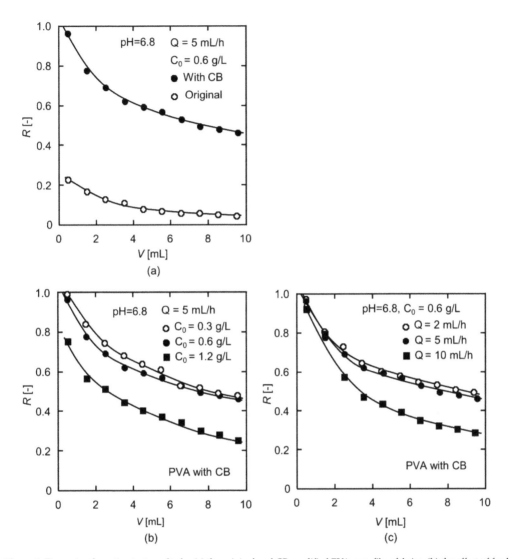

Figure 4. Dynamic adsorption test results for (**a**) the original and CB-modified PVA nanofiber fabrics, (**b**) the effect of feed concentrations on the CB-modified PVA nanofiber fabrics, and (**c**) the effect of permeation rates on the CB-modified PVA nanofiber fabrics.

Figure 4b presents the BHb rejection ratio of the CB-modified PVA nanofiber fabrics varied with permeation volume at different feed concentrations. For the highest feed concentration of 1.2 g/L, the contact probability for interaction between BHb molecules in the mobile phase and nanofiber surface was the highest, and the driving force was the largest. As a result, the BHb adsorption sites on the nanofiber surface were occupied sooner and the rejection ratio was markedly lower than that of the other two dilute solutions. For the solutions with concentrations of 0.3 and 0.6 g/L, the changes in the BHb rejection ratio were similar, but the rejection ratio values were higher at 0.3 g/L.

The influence of permeation rate on the BHb rejection ratio was examined, and the results are shown in Figure 4c. When the permeation rate was 10 mL/h, the BHb rejection ratio was dramatically reduced due to the insufficient retention time of BHb molecules in the pores of the nanofiber fabric. In contrast, the BHb rejection ratio decreased more slowly in the permeation experiments, with permeation rates of 2 and 5 mL/h. Therefore, it is necessary to allow for sufficient residence time in order to obtain the protein rejection effect. Additionally, attention needs to be paid to the balance between the productivity and rejection ratio during the dynamic adsorption operation.

3.4. Effect of pH on BHb and BSA Adsorption

Both BHb and BSA are homologous proteins derived from bovine blood, and their molecular weights are so close that they are difficult to separate with conventional filtration methods. Nevertheless, it is worth noting that the two proteins have different isoelectric points, 5.0 and 6.8, respectively. Proteins are amphoteric biomolecules because the amino acids that make up proteins contain both basic and acidic functional groups [43]. The protein surface charge depends on the pH value of the protein solution. At the isoelectric point, the electric charge of protein becomes zero. When the pH value of the protein solution is lower than the isoelectric point, the protein surface has a positive charge; in the contrary situation, the protein is negatively charged. Therefore, pH has a considerable influence on the adsorption capacity.

The pH effect on the adsorption capacity in a single component solution of BHb or BSA was studied by the static adsorption experiments. As BSA undergoes irreversible structural transitions in the range of pH < 4 [51], the experiments were performed at pH = 4–9. Figure 5 displays the adsorption amounts of the CB-modified PVA nanofiber fabrics towards BHb and BSA. Under different pH values, the adsorption capacities of BHb and BSA were different. The greater the difference between the adsorption amounts, the more likely it is to realize selective separation. The adsorption capacity of BHb was the maximum at its corresponding isoelectric point of 6.8. The BHb adsorption capacity decreased in varying degrees at other pH values. Similarly, the BSA adsorption capacity obtained its maximum at an isoelectric point of 5.0. At the isoelectric point, the repulsive force between protein molecules was minimal because the protein charge was zero. As a result, more protein molecules were arranged on the nanofiber surface during the affinity interaction with the CB molecules immobilized on the nanofiber fabrics. At pH = 6.8, the adsorption amounts of the CB-modified PVA nanofiber fabrics were 594 mg/g for BHb and 244 mg/g for BSA. The difference in the adsorption amounts was more extensive than that of other pH values, about 350 mg/g. In this case, pH = 6.8 was taken as the condition of the binary protein separation.

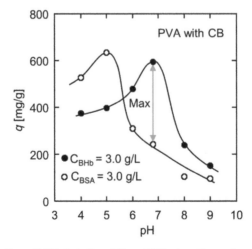

Figure 5. Effect of pH on BHb and BSA adsorption performances.

3.5. Selective Separation of Binary BHb–BSA Solution

Dynamic adsorption experiments of the CB-modified PVA nanofiber fabrics were conducted with binary BHb–BSA solutions. Figure 6 shows the variation in the rejection ratios for BHb and BSA versus the permeation volume. From the beginning of the permeation experiment, the rejection ratio for BHb was much higher than BSA, and the rejection ratios decreased gradually as the permeation volume increased. The selectivity factor S of BHb for the initial 1 mL of the permeate was about 3.01. The initial S seemed unsatisfactory due to a large number of vacant active sites on the nanofiber surface where BHb and BSA molecules were both able to attach at the beginning of the experiment. Similarly, a higher mass per unit area of the PVA nanofiber fabrics does not necessarily lead to an improvement in selectivity factor because the rejection ratio of both BHb and BSA will be maintained at a high level due to abundant adsorption sites. However, at an end permeation volume of 10 mL, the rejection ratios for BHb and BSA were 0.60 and 0.11, respectively, and S rose to 5.45. The dynamic adsorption operation was carried out only once, and it showed a good separation possibility. Higher selectivity factors are expected in multi-stage continuous dynamic adsorption operations, which needs to be investigated in future work.

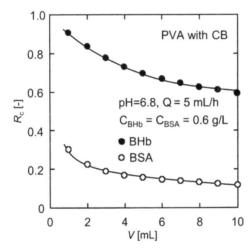

Figure 6. Selective separation results for the BHb–BSA binary solution.

The BHb molecules with zero charge occupied more active sites on the nanofiber surface. On the contrary, at pH = 6.8, BSA was negatively charged and had difficulty in close arrangements on the nanofiber surface due to the electrostatic repulsion between molecules, resulting in a weaker competition for active sites on the nanofiber surface compared to BHb. In addition, the size of the BHb molecule is 6.4 nm × 5.5 nm × 5 nm, with a spherical shape, while the size of the BSA molecule is 14 nm × 3.8 nm × 3.8 nm, with a shape similar to a prolate ellipsoid [52]. When BSA molecules flowed through the intricate pores inside the nanofiber fabrics, in order to obtain minimal interaction with the nanofiber surface, the ellipsoidal molecules might align their long axis parallel to the centerline of the nanofiber, resulting in a lower hydrodynamic hindrance than that of BHb molecules. In contrast, spherical BHb molecules were more easily attracted by the CB molecules immobilized on the nanofiber surface when passing through the internal pores of the nanofiber fabric.

3.6. Desorption and Reusability

It has been reported that increasing the solution ionic strength leads to a decrease in protein adsorption amount [38,53]. Salts in the adsorption medium can lead to coordination of the deprotonated sulfonic acid groups of CB molecule with sodium ions; on the other hand, the distortion of existing salt bridges contributes to low protein adsorption at high ionic strength [45]. In addition, moving the pH of the adsorption medium away from the protein's isoelectric point is also detrimental to adsorption. Thus, a phosphate buffer at pH = 10.0 containing 1.0 M NaCl was used as the eluent to desorb BHb and BSA from the protein adsorbed CB-modified PVA nanofiber fabrics. According to our previous analysis of protein concentrations at different pH values, there was no significant difference in absorbance at pH = 10 for the same concentration of protein solution while the absorbance severely decayed at pH ≤ 3. Therefore, BHb and BSA are more sensitive to acidity than alkalinity, and the eluent of pH = 10 does not cause structural damage to proteins. The protein-adsorbed nanofiber fabric was eluted with 20 mL eluent at a rate of 5 mL/h using the permeation apparatus, and more than 96% of the protein molecules were eluted off. The eluent can be desalinated by an ultrafiltration operation after pH adjustment, and then, the desorbed BHb and BSA can be concentrated for further recovery.

The adsorption–desorption was repeated for three cycles to evaluate the reusability of the CB-modified PVA nanofiber fabrics. Figure 7 displays the changes in rejection ratios for BHb and BSA with the permeation volume during the three adsorption cycles. The rejection ratios for both proteins declined slightly. However, the selectivity factor S maintained above 5 at the end permeate volume of 10 mL in all three cycles, with 5.36, 5.07 and 5.13, respectively. It is demonstrated that the CB-modified PVA nanofiber fabrics can be reused without losing activity easily.

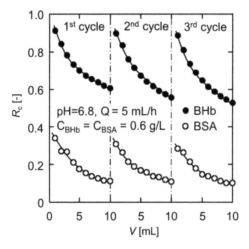

Figure 7. Reusability test of the CB-modified PVA nanofiber fabric for selective separation.

4. Conclusions

In this research, CB-modified PVA nanofiber fabrics were fabricated via immobilization of CB molecules on the nanofiber surface. The batch experiments showed that the modified nanofiber fabrics have a high adsorption capacity for BHb. The Langmuir isotherm was used to fit the experimental data, and the maximum adsorption capacity of the nanofiber fabric was calculated to be 686 mg/g. Then, dynamic experiments were performed to determine the effect of feed concentration and permeation rate on the BHb rejection ratio. In the single-protein static adsorption experiments, the pH impact on the adsorption of BHb and BSA was investigated separately, where the pH value with the largest difference in the adsorption amount was explored. Finally, a high selectivity factor of 5.45 for BHb was achieved during the dynamic adsorption of the BHb–BSA binary solution at pH = 6.8, corresponding to the isoelectric point of BHb. The selectivity factors for BHb were maintained above 5 in the three adsorption–elution cycles, indicating the reusability

Polymers **2021**, 13, 2313

of the nanofiber fabric. This work demonstrated the potential of the CB-modified PVA nanofiber fabric in the selective separation for proteins with similar sizes.

Author Contributions: Conceptualization, Y.M.; investigation, S.L.; resources, Y.M.; data curation, S.L.; writing—original draft preparation, S.L.; writing—review and editing, Y.M. All authors have read and agreed to the published version of the manuscript.

Funding: This research was funded by JSPS KAKENHI Grant No. JP20K05191.

Institutional Review Board Statement: Not applicable.

Informed Consent Statement: Not applicable.

Data Availability Statement: The data presented in this study are available from the corresponding author upon request.

Acknowledgments: The authors would like to acknowledge the support provided by JSPS KAKENHI Grant.

Conflicts of Interest: The authors declare no conflict of interest.

References

1. Watson, H. Biological membranes. *Essays Biochem.* **2015**, *59*, 43–70. [CrossRef] [PubMed]
2. Buehler, M.J.; Yung, Y.C. Deformation and failure of protein materials in physiologically extreme conditions and disease. *Nat. Mater.* **2009**, *8*, 175–188. [CrossRef] [PubMed]
3. Kuhlman, B.; Bradley, P. Advances in protein structure prediction and design. *Nat. Rev. Mol. Cell Biol.* **2019**, *20*, 681–697. [CrossRef] [PubMed]
4. Stefani, M.; Dobson, C.M. Protein aggregation and aggregate toxicity: New insights into protein folding, misfolding diseases and biological evolution. *J. Mol. Med.* **2003**, *81*, 678–699. [CrossRef] [PubMed]
5. Cournia, Z.; Allen, T.W.; Andricioaei, I.; Antonny, B.; Baum, D.; Brannigan, G.; Buchete, N.V.; Deckman, J.T.; Delemotte, L.; del Val, C.; et al. Membrane protein structure, function, and dynamics: A perspective from experiments and theory. *J. Membr. Biol.* **2015**, *248*, 611–640. [CrossRef] [PubMed]
6. Pace, C.N.; Grimsley, G.R.; Scholtz, J.M. Protein ionizable groups: pK values and their contribution to protein stability and solubility. *J. Biol. Chem.* **2009**, *284*, 13285–13289. [CrossRef]
7. Li, Y.; Stern, D.; Lock, L.L.; Mills, J.; Ou, S.H.; Morrow, M.; Xu, X.; Ghose, S.; Li, Z.J.; Cui, H. Emerging biomaterials for downstream manufacturing of therapeutic proteins. *Acta Biomater.* **2019**, *95*, 73–90. [CrossRef]
8. Kumar, P.; Sharma, N.; Ranjan, R.; Kumar, S.; Bhat, Z.F.; Jeong, D.K. Perspective of membrane technology in dairy industry: A review. *Asian Australas. J. Anim. Sci.* **2013**, *26*, 1347–1358. [CrossRef]
9. Kapoor, S.; Rafiq, A.; Sharma, S. Protein engineering and its applications in food industry. *Crit. Rev. Food Sci. Nutr.* **2017**, *57*, 2321–2329. [CrossRef]
10. Navarro, A.; Wu, H.S.; Wang, S.S. Engineering problems in protein crystallization. *Sep. Purif. Technol.* **2009**, *68*, 129–137. [CrossRef]
11. Dos Santos, R.; Carvalho, A.L.; Roque, A.C.A. Renaissance of protein crystallization and precipitation in biopharmaceuticals purification. *Biotechnol. Adv.* **2017**, *35*, 41–50. [CrossRef] [PubMed]
12. Tang, W.; Zhang, H.; Wang, L.; Qian, H.; Qi, X. Targeted separation of antibacterial peptide from protein hydrolysate of anchovy cooking wastewater by equilibrium dialysis. *Food Chem.* **2015**, *168*, 115–123. [CrossRef]
13. Chai, M.; Ye, Y.; Chen, V. Separation and concentration of milk proteins with a submerged membrane vibrational system. *J. Memb. Sci.* **2017**, *524*, 305–314. [CrossRef]
14. Hage, D.S.; Anguizola, J.A.; Bi, C.; Li, R.; Matsuda, R.; Papastavros, E.; Pfaunmiller, E.; Vargas, J.; Zheng, X. Pharmaceutical and biomedical applications of affinity chromatography: Recent trends and developments. *J. Pharm. Biomed. Anal.* **2012**, *69*, 93–105. [CrossRef] [PubMed]
15. Iritani, E.; Mukai, Y.; Tanaka, Y.; Murase, T. Flux decline behavior in dead-end microfiltration of protein solutions. *J. Memb. Sci.* **1995**, *103*, 181–191. [CrossRef]
16. Mukai, Y.; Iritani, E.; Murase, T. Fractionation characteristics of binary protein mixtures by ultrafiltration. *Sep. Sci. Technol.* **1998**, *33*, 169–185. [CrossRef]
17. Iritani, E.; Mukai, Y. Approach from physicochemical aspects in membrane filtration. *Korean J. Chem. Eng.* **1997**, *14*, 347–353. [CrossRef]
18. Casey, C.; Gallos, T.; Alekseev, Y.; Ayturk, E.; Pearl, S. Protein concentration with single-pass tangential flow filtration (SPTFF). *J. Memb. Sci.* **2011**, *384*, 82–88. [CrossRef]
19. Emin, C.; Kurnia, E.; Katalia, I.; Ulbricht, M. Polyarylsulfone-based blend ultrafiltration membranes with combined size and charge selectivity for protein separation. *Sep. Purif. Technol.* **2018**, *193*, 127–138. [CrossRef]

20. Huang, L.; Ye, H.; Yu, T.; Zhang, X.; Zhang, Y.; Zhao, L.; Xin, Q.; Wang, S.; Ding, X.; Li, H. Similarly sized protein separation of charge-selective ethylene-vinyl alcohol copolymer membrane by grafting dimethylaminoethyl methacrylate. *J. Appl. Polym. Sci.* **2018**, *135*, 1–8. [CrossRef]
21. Nadar, S.S.; Pawar, R.G.; Rathod, V.K. Recent advances in enzyme extraction strategies: A comprehensive review. *Int. J. Biol. Macromol.* **2017**, *101*, 931–957. [CrossRef]
22. Huang, L.B.; Xu, W.; Zhao, C.; Zhang, Y.L.; Yung, K.L.; Diao, D.; Fung, K.H.; Hao, J. Multifunctional water drop energy harvesting and human motion sensor based on flexible dual-mode nanogenerator incorporated with polymer nanotubes. *ACS Appl. Mater. Interfaces* **2020**, *12*, 24030–24038. [CrossRef]
23. Huang, L.B.; Xu, W.; Tian, W.; Han, J.C.; Zhao, C.H.; Wu, H.L.; Hao, J. Ultrasonic-assisted ultrafast fabrication of polymer nanowires for high performance triboelectric nanogenerators. *Nano Energy* **2020**, *71*, 104593. [CrossRef]
24. Huang, L.B.; Xu, W.; Hao, J. Energy device applications of synthesized 1D polymer nanomaterials. *Small* **2017**, *13*, 1701820. [CrossRef]
25. Lee, H.; Yamaguchi, K.; Nagaishi, T.; Murai, M.; Kim, M.; Wei, K.; Zhang, K.Q.; Kim, I.S. Enhancement of mechanical properties of polymeric nanofibers by controlling crystallization behavior using a simple freezing/thawing process. *RSC Adv.* **2017**, *7*, 43994–44000. [CrossRef]
26. Mamun, A.; Blachowicz, T.; Sabantina, L. Electrospun nanofiber mats for filtering applications—Technology, structure and materials. *Polymers* **2021**, *13*, 1368. [CrossRef] [PubMed]
27. Yoon, K.; Hsiao, B.S.; Chu, B. Functional nanofibers for environmental applications. *J. Mater. Chem.* **2008**, *18*, 5326–5334. [CrossRef]
28. Choi, H.Y.; Bae, J.H.; Hasegawa, Y.; An, S.; Kim, I.S.; Lee, H.; Kim, M. Thiol-functionalized cellulose nanofiber membranes for the effective adsorption of heavy metal ions in water. *Carbohydr. Polym.* **2020**, *234*, 115881. [CrossRef] [PubMed]
29. Aliabadi, M.; Irani, M.; Ismaeili, J.; Piri, H.; Parnian, M.J. Electrospun nanofiber membrane of PEO/Chitosan for the adsorption of nickel, cadmium, lead and copper ions from aqueous solution. *Chem. Eng. J.* **2013**, *220*, 237–243. [CrossRef]
30. Pant, B.; Barakat, N.A.M.; Pant, H.R.; Park, M.; Saud, P.S.; Kim, J.W.; Kim, H.Y. Synthesis and photocatalytic activities of CdS/TiO2 nanoparticles supported on carbon nanofibers for high efficient adsorption and simultaneous decomposition of organic dyes. *J. Colloid Interface Sci.* **2014**, *434*, 159–166. [CrossRef]
31. Ke, X.; Huang, Y.; Dargaville, T.R.; Fan, Y.; Cui, Z.; Zhu, H. Modified alumina nanofiber membranes for protein separation. *Sep. Purif. Technol.* **2013**, *120*, 239–244. [CrossRef]
32. Lan, T.; Shao, Z.Q.; Wang, J.Q.; Gu, M.J. Fabrication of hydroxyapatite nanoparticles decorated cellulose triacetate nanofibers for protein adsorption by coaxial electrospinning. *Chem. Eng. J.* **2015**, *260*, 818–825. [CrossRef]
33. Fu, Q.; Duan, C.; Yan, Z.; Si, Y.; Liu, L.; Yu, J.; Ding, B. Electrospun nanofibrous composite materials: A versatile platform for high efficiency protein adsorption and separation. *Compos. Commun.* **2018**, *8*, 92–100. [CrossRef]
34. Gibson, P.; Schreuder-Gibson, H.; Rivin, D. Transport properties of porous membranes based on electrospun nanofibers. *Colloids Surf. A Physicochem. Eng. Asp.* **2001**, *187*, 469–481. [CrossRef]
35. Duan, C.; Fu, Q.; Si, Y.; Liu, L.; Yin, X.; Ji, F.; Yu, J.; Ding, B. Electrospun regenerated cellulose nanofiber based metal-chelating affinity membranes for protein adsorption. *Compos. Commun.* **2018**, *10*, 168–174. [CrossRef]
36. Show, P.L.; Ooi, C.W.; Lee, X.J.; Yang, C.L.; Liu, B.L.; Chang, Y.K. Batch and dynamic adsorption of lysozyme from chicken egg white on dye-affinity nanofiber membranes modified by ethylene diamine and chitosan. *Int. J. Biol. Macromol.* **2020**, *162*, 1711–1724. [CrossRef] [PubMed]
37. Ng, I.S.; Song, C.P.; Ooi, C.W.; Tey, B.T.; Lee, Y.H.; Chang, Y.K. Purification of lysozyme from chicken egg white using nanofiber membrane immobilized with Reactive Orange 4 dye. *Int. J. Biol. Macromol.* **2019**, *134*, 458–468. [CrossRef]
38. Liu, S.; Mukai, Y. Dynamic adsorption behaviors of protein on Cibacron Blue-modified PVA nanofiber fabrics. *J. Text. Eng.* **2021**, *67*, 1–11. [CrossRef]
39. Liu, S.; Sumi, T.; Mukai, Y. Development of Cibacron Blue-enhanced affinity nanofiber fabric for protein adsorption. *J. Fiber Sci. Technol.* **2020**, *76*, 327–334. [CrossRef]
40. Das, S.; Bora, N.; Rohman, M.A.; Sharma, R.; Jha, A.N.; Singha Roy, A. Molecular recognition of bio-active flavonoids quercetin and rutin by bovine hemoglobin: An overview of the binding mechanism, thermodynamics and structural aspects through multi-spectroscopic and molecular dynamics simulation studies. *Phys. Chem. Chem. Phys.* **2018**, *20*, 21668–21684. [CrossRef]
41. Carter, D.C.; Ho, J.X. Structure of serum albumin. *Adv. Protein Chem.* **1994**, *45*, 153–176. [CrossRef] [PubMed]
42. Reddy, S.M.; Sette, G.; Phan, Q. Electrochemical probing of selective haemoglobin binding in hydrogel-based molecularly imprinted polymers. *Electrochim. Acta* **2011**, *56*, 9203–9208. [CrossRef]
43. Peters, T. Serum albumin. *Adv. Protein Chem.* **1985**, *37*, 161–245. [CrossRef] [PubMed]
44. Ruckenstein, E.; Zeng, X. Albumin separation with Cibacron Blue carrying macroporous chitosan and chitin affinity membranes. *J. Memb. Sci.* **1998**, *142*, 13–26. [CrossRef]
45. Yavuz, H.; Duru, E.; Genç, Ö.; Denizli, A. Cibacron Blue F3GA incorporated poly(methylmethacrylate) beads for albumin adsorption in batch system. *Colloids Surf. A Physicochem. Eng. Asp.* **2003**, *223*, 185–193. [CrossRef]
46. Liang-Schenkelberg, J.; Fieg, G.; Waluga, T. Molecular insight into affinity interaction between Cibacron Blue and proteins. *Ind. Eng. Chem. Res.* **2017**, *56*, 9691–9697. [CrossRef]
47. Sharma, S.; Agarwal, G.P. Interactions of proteins with immobilized metal ions: A comparative analysis using various isotherm models. *Anal. Biochem.* **2001**, *288*, 126–140. [CrossRef]

48. Lan, T.; Shao, Z.Q.; Gu, M.J.; Zhou, Z.W.; Wang, Y.L.; Wang, W.J.; Wang, F.J.; Wang, J.Q. Electrospun nanofibrous cellulose diacetate nitrate membrane for protein separation. *J. Memb. Sci.* **2015**, *489*, 204–211. [CrossRef]

49. Zhang, M.; Wang, Y.; Zhang, Y.; Ding, L.; Zheng, J.; Xu, J. Preparation of magnetic carbon nanotubes with hierarchical copper silicate nanostructure for efficient adsorption and removal of hemoglobin. *Appl. Surf. Sci.* **2016**, *375*, 154–161. [CrossRef]

50. Wang, J.; Tan, S.; Liang, Q.; Guan, H.; Han, Q.; Ding, M. Selective separation of bovine hemoglobin using magnetic mesoporous rare-earth silicate microspheres. *Talanta* **2019**, *204*, 792–801. [CrossRef]

51. Song, H.; Yang, C.; Yohannes, A.; Yao, S. Acidic ionic liquid modified silica gel for adsorption and separation of bovine serum albumin (BSA). *RSC Adv.* **2016**, *6*, 107452–107462. [CrossRef]

52. Qiu, X.; Yu, H.; Karunakaran, M.; Pradeep, N.; Nunes, S.P.; Peinemann, K.V. Selective separation of similarly sized proteins with tunable nanoporous block copolymer membranes. *ACS Nano* **2013**, *7*, 768–776. [CrossRef] [PubMed]

53. Tuzmen, N.; Kalburcu, T.; Uygun, D.A.; Akgol, S.; Denizli, A. A novel affinity disks for bovine serum albumin purification. *Appl. Biochem. Biotechnol.* **2015**, *175*, 454–468. [CrossRef] [PubMed]

Article

Nanofiber-Mâché Hollow Ball Mimicking the Three-Dimensional Structure of a Cyst

Wan-Ying Huang [1], Norichika Hashimoto [2], Ryuhei Kitai [3], Shin-ichiro Suye [1,4,5] and Satoshi Fujita [1,4,5,*]

1   Department of Advanced Interdisciplinary Science and Technology, Graduate School of Engineering,
    University of Fukui, 3-9-1 Bukyo, Fukui-shi 910-8507, Japan; ivy@biofiber-fukui.com (W.-Y.H.);
    suyeb10@u-fukui.ac.jp (S.-i.S.)
2   Department of Neurosurgery, Fukui Health Sciences University, 55-13-1 Egami, Fukui-shi 910-3190, Japan;
    hashi19721223@gmail.com
3   Department of Neurosurgery, Kaga Medical Center, Ri 36 Sakumi, Kaga-shi 922-8522, Japan; rkitai@gmail.com
4   Department of Frontier Fiber Technology and Science, University of Fukui,
    3-9-1 Bukyo, Fukui-shi 910-8507, Japan
5   Organization for Life Science Advancement Programs, University of Fukui,
    3-9-1 Bukyo, Fukui-shi 910-8507, Japan
*   Correspondence: fujitas@u-fukui.ac.jp; Tel.: +81-776-27-9969

Citation: Huang, W.-Y.; Hashimoto, N.; Kitai, R.; Suye, S.-i.; Fujita, S. Nanofiber-Mâché Hollow Ball Mimicking the Three-Dimensional Structure of a Cyst. *Polymers* **2021**, *13*, 2273. https://doi.org/10.3390/polym13142273

Academic Editors: Ick-Soo Kim, Sana Ullah and Motahira Hashmi

Received: 15 June 2021
Accepted: 8 July 2021
Published: 11 July 2021

Publisher's Note: MDPI stays neutral with regard to jurisdictional claims in published maps and institutional affiliations.

Abstract: The occasional malignant transformation of intracranial epidermoid cysts into squamous cell carcinomas remains poorly understood; the development of an in vitro cyst model is urgently needed. For this purpose, we designed a hollow nanofiber sphere, the "nanofiber-mâché ball." This hollow structure was fabricated by electrospinning nanofiber onto alginate hydrogel beads followed by dissolving the beads. A ball with approximately 230 mm^3 inner volume provided a fibrous geometry mimicking the topography of the extracellular matrix. Two ducts located on opposite sides provided a route to exchange nutrients and waste. This resulted in a concentration gradient that induced oriented migration, in which seeded cells adhered randomly to the inner surface, formed a highly oriented structure, and then secreted a dense web of collagen fibrils. Circumferentially aligned fibers on the internal interface between the duct and hollow ball inhibited cells from migrating out of the interior, similar to a fish bottle trap. This structure helped to form an adepithelial layer on the inner surface. The novel nanofiber-mâché technique, using a millimeter-sized hollow fibrous scaffold, is excellently suited to investigating cyst physiology.

Keywords: electrospinning; nanofiber; hollow ball; alginate; tissue engineering; 3D structure

1. Introduction

Cysts are spherical sac-like multicellular structures with fluid and semisolid matter inside. Cysts occur in every kind of tissue; they are usually harmless and heal spontaneously, but occasionally transform into malignant tumors. For example, an intracranial epidermoid cyst sometimes undergoes a malignant transformation into an intracranial squamous-cell carcinoma, which has a poor prognosis. Treatment of malignancies is difficult, especially if the brain stem is extensively involved in the cyst [1,2]. Most investigations of cyst biology have involved in vivo experimental models; elucidating cellular behaviors in the cyst in vitro has proved difficult. Previously reported methods of preparing three-dimensional (3D) cellular structures include the hanging drop method [3], the formation of spheroids by cell non-adhesive multi-plate [4] or micropatterned-substrate [5] methods, and the use of cell sheets [6], bioprinting [7], and hydrogels [8] for forming 3D cell aggregates. However, cell aggregates prepared by these methods are cell-filled structures; it has been impossible to form a hollow structure such as a cyst.

The approach introduced in this study was inspired by the fabrication of papier-mâché dolls from balloons and paper. Our model cyst is a controlled nano-structure, the "nanofiber-mâché ball"—a hollow nanofiber sphere made of biodegradable polyester. We have focused on electrospinning as a promising technique to fabricate geometrically controlled nanofibers [9–11]. Electrospun nanofibers with high orientation were deposited on the surface of hydrogel beads of alginate, a natural hydrophilic and anionic polysaccharide commonly obtained from brown algae that exhibits good biocompatibility and biodegradability [12,13]. Alginate hydrogel has been extensively investigated for biomedical applications such as cell scaffolding and drug release [10,14–16]. Alginate can easily form a hydrogel in the presence of divalent cations such as calcium, while cross-linked alginic acid can be chelated by ethylenediaminetetraacetic acid (EDTA) for easy dissolution [17]. We chose poly(3-hydroxybutyrate-*co*-3-hydroxyhexanoate) (PHBH), a biocompatible and flexible polyester, as the matrix fiber [18–20]. We used electrospinning to fabricate nanofiber with a specific topography that mimics the extracellular matrix (ECM) for controlling cell fate. To our knowledge, this is the first report of fabrication of a millimeter-sized and geometrically controlled cyst-like structure.

2. Materials and Methods

2.1. Materials

PHBH (3HH = 10.6 mol%) was provided by Kaneka (Tokyo, Japan); sodium alginate (500 cps) was obtained from Nacalai Tesque (Kyoto, Japan); chloroform (CHCl$_3$) and calcium chloride (CaCl$_2$) were obtained from FUJIFILM Wako Pure Chemical (Osaka, Japan); EDTA was obtained from Thermo Fisher Scientific (Waltham, MA, USA). All other chemicals and reagents were of analytical grade and were used without further purification.

2.2. Fabrication of Nanofiber-Mâché Ball

The fabrication process is shown in Scheme 1. Alginate hydrogel beads (3 mm in diameter) were obtained by dropping 5 w/v% sodium alginate intermittently into 1 w/v% CaCl$_2$. An aluminum wire (0.5 mm in diameter) penetrated the alginate beads and was then set up along the rotating axis of the electrospinning collector. PHBH solution (15 w/v% in CHCl$_3$) was electrospun onto the alginate gel beads at 0.1 mL/h flow rate with 29 kV and a needle-tip–to–collector distance of 10 cm. The density and orientation were controlled by the spinning time and rotation speed of the collector, respectively. The innermost layer consisted of random fibers electrospun at the rotation speed of 100 rpm for 45 min; the intermediate layer of weakly aligned fibers electrospun at 500 rpm for 25 min; and the outermost layer of strongly aligned fibers electrospun at 1000 rpm for 25 min. After electrospinning, the alginate hydrogel gelled with calcium ions was washed with 10 mM EDTA solution six times for 10 min and kept in EDTA solution overnight to chelate the calcium ions. After washing and drying, the nanofiber-mâché ball was used in cell culture.

2.3. Morphological Observation

Morphological observation of the nanofiber-mâché structure was carried out using digital microscopy (VHX-2000, KEYENCE, Osaka, Japan) and scanning electron microscopy (SEM; JCM-6000Plus, JEOL, Tokyo, Japan). The fiber samples were immersed in liquid nitrogen and cut into several small pieces in a frozen condition.

For SEM observation, cells were fixed with fresh 4% paraformaldehyde for 30 min, washed, dehydrated with sequentially diluted ethanol, and then transferred into t-butyl alcohol for 2 h. Both fiber samples and cell samples were lyophilized with a freeze-dryer. The samples were sputtered with Os for 10 s using an ion sputter (HPC-1SW, Vacuum Device Inc., Ibaragi, Japan) and observed using SEM at an accelerating voltage of 15 kV.

To measure the diameter of nanofiber and collagen fibrils, 100 randomly selected fibers were counted using image-processing software (Fiji.sc; Ver 2.0.0, ImageJ, NIH, USA). The nanofiber orientation was quantified by the second-order parameter S, defined by

$$S = 2 < \cos^2 \theta > -1 = < \cos 2\theta > \tag{1}$$

where θ represents the orientation angle and $< \cos 2\theta >$ is the average of $\cos 2\theta$ [21,22].

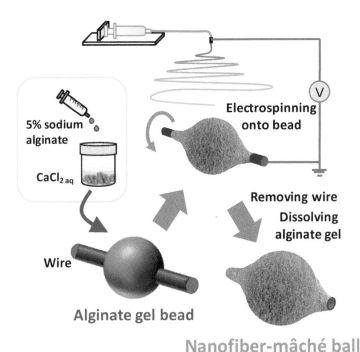

Scheme 1. Conceptual illustration of the fabrication of nanofiber-mâché ball based on alginate gel beads.

2.4. Surface Characterization

Attenuated total reflectance-Fourier transform infrared (ATR-FTIR) spectrometry measurement was performed to verify the composition of nanofibers on a Nicolet 6700 system (Thermo Scientific, Waltham, MA, USA), with the wave number range from 500 cm^{-1} to 4000 cm^{-1} at a resolution of 4 cm^{-1} with an average of 64 scans. The hydrophilicity of the surface of nanofiber electrospun on a cover slip was measured using a water contact angle meter (DSA25E, KRÜSS, Hamburg, Germany).

2.5. Cell Culture with Nanofiber-Mâché Ball

To improve its surface hydrophilicity, the nanofiber-mâché ball was treated with oxygen plasma (40 kHz/100 W, 0.1 MPa, 30 s; Femto 9, Diener electronic, Ebhausen, Germany) and washed twice with 30% EtOH and PBS, then coated with 10 µg·mL^{-1} of fibronectin for 30 min at 37 °C. U-251 MG cells, derived from human glioblastoma astrocytoma, were cultured in high glucose Dulbecco's modified Eagle's medium (DMEM; Fujifilm Wako Pure Chemical Corporation, Osaka, Japan) supplemented with 10% fetal bovine serum (FBS; JRH Biosciences, Lenexa, KS, USA) and 1% Penicillin-Streptomycin solution (Fujifilm Wako Pure Chemical Corporation). Cell suspension, mixed with 3.75% methylcellulose (#400, Nacalai Tesque, Japan), was injected into the nanofiber-mâché ball at a density of 1×10^4 cells/µL and incubated in a humidified atmosphere with 5% CO_2 at 37 °C for 72 h.

2.6. Statistical Analysis

All values were expressed as mean and standard deviation. Student's *t*-test was performed and $p < 0.05$ was considered statistically significant.

3. Results

3.1. Morphology of Nanofiber-Mâché Ball

PHBH fiber was electrospun on an alginate bead fixed on a wire (Figure 1a). To reveal the outer and inner structural morphology of the fabricated hollow nanofiber-mâché balls after the washing process, two semi-spheres were observed using digital microscopy (Figure 1b,c). The actual inner and outer diameters of a nanofiber-mâché ball were found to be approximately 3.8 and 4.1 mm, respectively. The interior of the ball, corresponding to the microenvironment where cells reside in tissues, had a volume of 230 mm^3. The two ducts formed after the core aluminum wire was removed, located at opposite sides of the hollow ball, had inner diameters of approximately 0.58 mm; they can be used to access the interior for injection of cell suspension. Fine structures were observed in detail by SEM. In a sectional view of the nanofiber-mâché ball, it was evident that the inner and outer structures were distinct. The inner fibers were spun randomly to form a smooth surface (Figure 1d), whereas the outer fibers were spun along the circumferential direction (Figure 1e). Longitudinal-section observation of a nanofiber-mâché ball and duct showed details of the interface between the central sphere and duct on both sides (Figure 1f, arrowhead). The cross-section of the wall showed the geometric difference between the inner and outer surfaces (Figure 1g). The thickness of the inner layer was approximately 60.5 μm, while the outer layer was 53.9 μm in thickness. The total thickness of the wall was 114.4 μm.

3.2. Dimension and Geometry of Nanofiber

To reveal the dimension and geometry of electrospun nanofiber, the inner and outer surfaces were observed under high magnification (Figure 2a,b). The fibers on the inside were random; they were fused to each other as if they were pressed, because they were spun on the surface of alginate gel beads. The fibers were densely packed and adhered to each other, especially in the inner random region. Therefore, there was not enough space for cells to penetrate through the wall, although it was porous. This indicates that nutrients and oxygen can be supplied from the outside, but cells can migrate only through the duct.

The fiber diameter and secondary orientation parameters were measured from SEM images. The inner fibers were 1.70 ± 0.41 μm in diameter and had orientation parameter $S = 0.11 ± 0.06$ (Figure 2c,e). On the other hand, the fiber diameter and orientation parameter of the outer fibers were 1.76 ± 0.57 μm and 0.51 ± 0.07, respectively (Figure 2d,f). The size of the porous gap between the fibers was as same as the fiber diameter; we assumed that cells could move freely in such a space. We found within that fiber diameter no statistically significant changes when we altered the nanotopography of the nanofiber-mâché ball. It should be relatively easy to obtain balls of various types: orientation parameter is controllable by manipulating the size of the alginate beads and the rotation speed while electrospinning.

3.3. Characterization of Nanofiber Surface

ATR-FTIR spectra of PHBH, alginate, and the inner and outer surfaces of the nanofiber-mâché ball were measured (Figure 3). A specific peak of PHBH (1720 cm^{-1}, C = O stretching) was observed in both inner and outer surfaces of the nanofiber-mâché ball, but no peak attributed to alginate was observed, which indicates that there was no residual alginate in the nanofiber-mâché ball after the dissolving process of alginate.

To show the improvement of surface hydrophilicity by oxygen plasma treatment, water contact angles of electrospun PHBH fiber sheets were measured by a sessile drop method (Figure 4). The water contact angles before and after plasma treatment were 102° ± 1° and 31.5° ± 1.5°, respectively. This result shows that oxygen plasma treatment hydrophilized PHBH, a hydrophobic polymer.

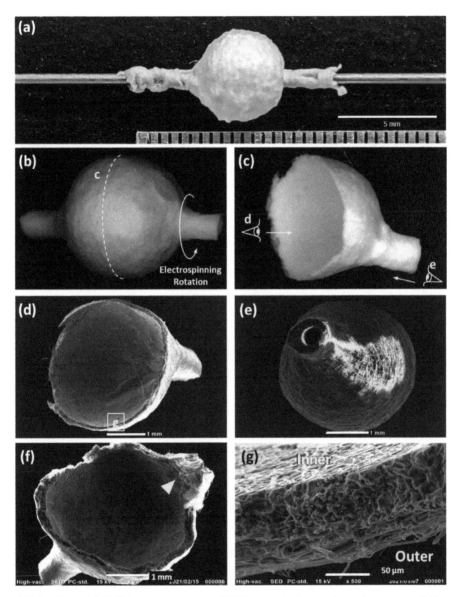

Figure 1. (**a**) Nanofiber-mâché ball electrospun on a wire. (**b**) Macroscopic pictures of overall structure and (**c**) nanofiber-mâché semi-sphere. (**d–g**) SEM images showing morphology of nanofiber-mâché ball. (**d**) Front view of semi-sphere, showing cup structure. (**e**) Back view of semi-sphere, showing a cell-seeding hole. (**f**) Longitudinal section of nanofiber-mâché ball and duct. (Arrowhead: interface between duct and central sphere.) (**g**) High-magnification cross-sectional view of nanofiber-mâché ball. The different fiber geometries and degrees of alignment on the inner and outer surfaces was observed.

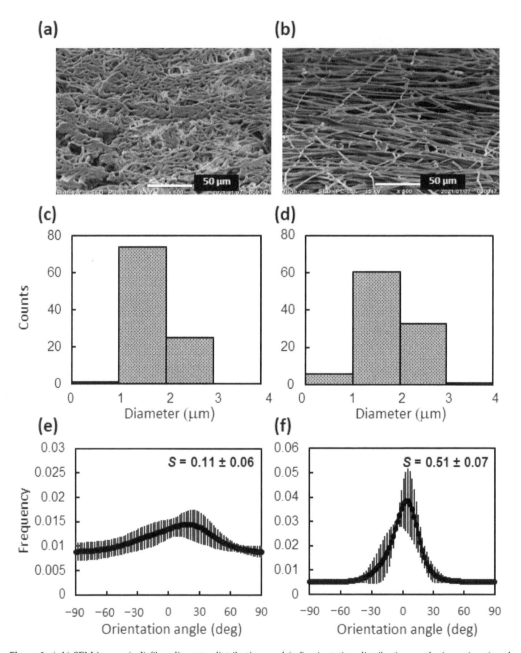

Figure 2. (**a,b**) SEM image, (**c,d**) fiber-diameter distribution, and (**e,f**) orientation distribution on the inner (**a,c,e**) and outer surface (**b,d,f**) of a nanofiber-mâché ball. The orientations of the inner (random) and outer (aligned) fibers were significantly different.

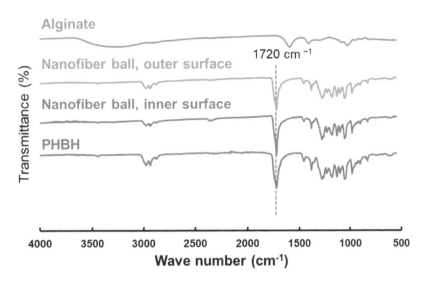

Figure 3. ATR-FTIR spectra of the alginate, inner, and outer surface of nanofiber-mâché ball, and PHBH.

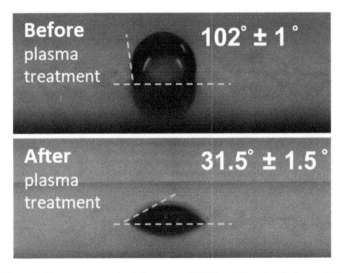

Figure 4. Water contact angle of electrospun PHBH nanofiber sheet before and after oxygen plasma treatment.

3.4. Cell Adhesion

While investigating the feasibility of cell culture in the nanofiber-mâché ball, we observed that U-251 MG cells (inoculated as a suspension mixed with methylcellulose medium) adhered to the whole fibronectin-coated inner surface of the ball (Figure 5a), the viscous medium having prevented them from settling. Interestingly, although the cells spread on the nanofibers, few penetrated beneath the tightly packed fibers (Figure 5b,c), so an adepithelial layer formed on the inner surface. This result indicates that the nanofiber mâché ball would provide a good microenvironment for cell adhesion.

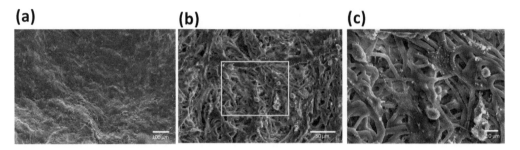

Figure 5. SEM images of U-251 MG culture on fibronectin coating nanofiber-mâché ball for 72 h. (**a**) Cells adhered to the inner surface, where the fibers were randomly oriented, at low magnification observation. (**b**) The adhesivity of uniform cells onto the inner surface (**c**) contributed to forming an adepithelial layer, at two magnifications. (Yellow in false-color image: cells).

3.5. Cell Orientation

To investigate the production of ECM in long-term cell culture, U-251 MG cells were inoculated into a ball that had been treated with oxygen plasma but not coated with fibronectin. In the seeding, cells were suspended in the medium without methylcellulose. After three weeks of culture, the ball was cut for SEM observation (Figure 6a). Cells had proliferated to cover the inner surface confluently (Figure 6b). As shown in Figure 6c, the cells formed highly orientated structures and were elongated in the radial direction toward the center of the duct. The reason that the cells aligned even though the inner surface consisted of random-spun fibers was the difference in the local concentration of nutrients and oxygen in the nanofiber-mâché ball. Each ball had two opposed ducts allowing access by the external medium with its nutrients and oxygen. The gradient in the concentration is a possible attractant for cell spreading and migration.

However, the cells near the periphery of the duct appeared to be randomly directed rather than aligned towards the duct (Figure 6d, arrowheads). It was also found that, at the sphere-duct interface itself, the fibers were aligned along the circumference of the duct (Figure 6d, curved arrow). This orientation of fibers was the result of electrospinning directly onto a conductive wire, not an alginate bead. When cells migrating out to the duct from the inner surface reached the transitional region, where the interfacial fibers aligned in the direction of the circumference, they were trapped and spread over randomly because the fiber orientation was perpendicular to the direction of migration. This fiber structure would work as a barrier to inhibit cells from migrating toward the outside, such as the one-way structure of a fish bottle trap.

The cross-section of the walls after long-term culture revealed the extent of cellular invasion (Figure 6e). Although cells in a layer covered the inner surface, they did not penetrate into the fibers deeply. The walls were non-porous and dense enough to inhibit cell migration. This observation indicates that the nanofiber-mâché ball could provide an adequate cell-adhesive surface.

3.6. ECM Production

High-magnification observation also revealed the production of a dense network of extracellular matrix structures (Figure 7a). Although the pre-fibronectin coating was not carried out in this culture, thin fibers with diameters 225 ± 136 nm were observed; they were much thinner than the PHBH scaffold and the cell itself (Figure 7b). Because they were similar in size to collagen fibrils, we consider them to be ECM fibers produced by U-251 MG cells.

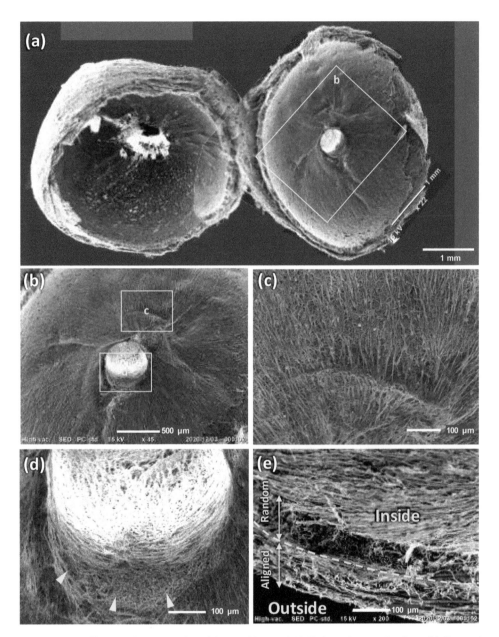

Figure 6. U-251 MG adhered to the inner surface of the nanofiber-mâché ball after a three-week culture. (**a**) Observation of the structure of the hollow nanofiber-mâché ball cut in half. (**b**) A cell layer tightly covered the inner surface after long-term culture. (**c**) The gradient in the concentration of the medium promoted cell alignment towards the duct exit. (**d**) Enlarged view of duct entrance. Cells oriented along aligned in the circumferential direction between the central sphere and ducts, but the periphery of the duct showed random towards the duct. (Curved arrow: circumferential fiber at the interface; arrowheads: fiber orientations on the periphery of the duct). (**e**) Invasion of cells (cross-sectional view). A layer of cells was covering the inner surface showed no obvious invasion into fibers.

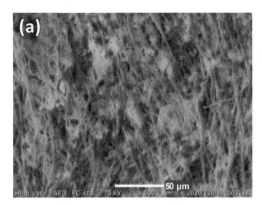
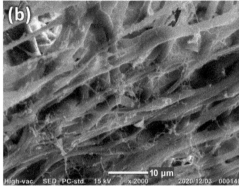

Figure 7. (a) High-density extracellular matrix (b) secreted web of collagen fibrils from U251 MG cells in a nanofiber-mâché ball.

4. Discussion

The 3D structures of intracranial epidermoid cysts consist of a thin outer layer of fluid-filled epithelial cells, keratin, and cholesterol. Establishing controllable-scale in vitro models of cysts will make our knowledge of pathogenesis more sophisticated, with prospective implications for optimizing the diagnosis of craniofacial dermoid cysts. There are two types of cutaneous cysts: dermoid, with epithelial cell walls containing either stratified or unstratified squamous epithelium, and periapical, encompassed by connective or granulation tissue. The former are often detected during infancy and early childhood, superficially or in the anterior orbit [23]. Their clinical severity, including the possibility of rupture and intracranial extension, may be directly affected by size and histologic differences [24]. In some clinical cases, craniofacial dermoid cysts of the frontotemporal showed subcutaneous sizes smaller than 15 mm × 15 mm in children [25]. At the other extreme, colloid cysts due to certain anatomical transmutations of the forniceal structures may have abnormally large sizes (>30 mm), and in an advanced stage of growth, a retro- or post-foraminal colloid cyst may separate from the forniceal structures if it is situated in the cavum septi pellucidi or vergae [26]. To determine the origin of cysts precisely from their sizes and histologic differences and predict the malignant possibility of lesions is a great challenge.

The nanofiber-mâché ball we developed is helpful for mimicking biological processes, in which the geometry and dimensions of 3D structures play key regulatory roles. The fiber orientation around the duct traps the cells inside the ball and prevents them from migrating out (Figure 6); the duct allows only fluid to pass between the interior and exterior. The narrow diameter of the duct also might induce small vortical flow, with cells orienting themselves along the direction of flow during culture [27]. This structure is a valid in vitro model of a fluid-filled cyst that is able to support tissues in vivo for oxygenation, nutrients, and waste removal.

The nanofiber-mâché ball with complex topography also provides efficient production of ECM network structure inside after long-term cell culture (Figure 7). The ECM is a composite network of the secreted products of resident cells containing structural and functional proteins that assemble in unique tissue-specific architecture; it serves as the scaffolding for tissues and organs in the body [28]. The microenvironment for tissue is supplied by ECM [29], including structural proteins such as collagen, cell-adhesion molecules such as fibronectin and laminin, and proteoglycans, which provide cells with a reservoir for biological signal molecules. Cells grow naturally in this 3D microenvironment that allows cell-cell and cell-matrix interaction and provides efficient biological signaling. Cell culture on a 3D model gives rise to cell interactions because its spatial packing influences a series of cellular behaviors, including cell morphology, proliferation, differentiation, the expression of genes and proteins, and cellular responses to extrinsic stimulation [30]. It can be used for the further investigation of the biological role of ECM in cysts.

Our model also has the potential to develop the optimal therapeutic strategy of a giant thrombotic aneurysm, which is a type of proliferative cerebral aneurysms because it absorbs nutrients from the outer blood vessel wall even if coil embolization or parent vessel occlusion treatment was performed at the inside of the aneurysm [31,32]. The nanofiber-mâché ball model can be used for the analysis of the growing mechanism of thrombosed aneurysms characterized by organized intraluminal thrombus and solid mass. Moreover, we expect that the bioprinting of alginate hydrogel has been rapidly advancing and it would be also applicable to prepare templates for nanofiber-mâché mimicking other tissues and organs.

5. Conclusions

We have successfully created a novel 3D hollow nanofiber-mâché ball by dissolving alginate beads. The hollow spaces left by the beads and the anisotropy of the nanofibers mimic the environment of cysts for cell-culture models. PHBH hollow nanofiber-mâché balls provide an in vitro culture system for the study of cell-cell and cell-ECM interactions. Our experimental model facilitates the analysis of biological behavior inside the cyst and may eventually shed light on phenomena that have not been fully elucidated clinically, such as the malignant transformation from epidermoid cysts to squamous cell carcinomas. Although further research is still needed, this preliminary study suggests great potential for medical applications.

Author Contributions: Conceptualization, N.H., R.K. and S.F.; methodology, S.F.; validation, W.-Y.H.; investigation, S.F.; data curation, W.-Y.H.; writing—original draft preparation, W.-Y.H.; writing—review and editing, N.H., R.K., S.-i.S. and S.F.; visualization, W.-Y.H.; supervision, S.F. All authors have read and agreed to the published version of the manuscript.

Funding: This research received no external funding.

Institutional Review Board Statement: Not applicable.

Informed Consent Statement: Not applicable.

Data Availability Statement: The raw/processed data required to reproduce these findings cannot be shared at this time due to legal reasons.

Acknowledgments: We acknowledge Terumo Life Science Foundation and the Grant-in-Aid for Scientific Research (C) (19K09502) from the Japan Society for the Promotion of Science. We acknowledge Hideyuki Uematsu for microscopic observation, and Kenji Hisada for contact angle measurement.

Conflicts of Interest: The authors declare no conflict of interest.

References

1. Faltaous, A.A.; Leigh, E.C.; Ray, P.; Wolbert, T.T. A Rare Transformation of Epidermoid Cyst into Squamous Cell Carcinoma: A Case Report with Literature Review. *Am. J. Case Rep.* **2019**, *20*, 1141–1143. [CrossRef] [PubMed]
2. Osborn, A.G.; Preece, M.T. Intracranial Cysts: Radiologic-Pathologic Correlation and Imaging Approach. *Radiology* **2006**, *239*, 650–664. [CrossRef] [PubMed]
3. Khoshnood, N.; Zamanian, A. A comprehensive review on scaffold-free bioinks for bioprinting. *Bioprinting* **2020**, *19*, e00088. [CrossRef]
4. Cho, M.S.; Kim, S.J.; Ku, S.Y.; Park, J.H.; Lee, H.; Yoo, D.H.; Park, U.C.; Song, S.A.; Choi, Y.M.; Yu, H.G. Generation of retinal pigment epithelial cells from human embryonic stem cell-derived spherical neural masses. *Stem Cell Res.* **2012**, *9*, 101–109. [CrossRef]
5. Théry, M. Micropatterning as a tool to decipher cell morphogenesis and functions. *J. Cell Sci.* **2010**, *123*, 4201–4213. [CrossRef] [PubMed]
6. Imai, M.; Furusawa, K.; Mizutani, T.; Kawabata, K.; Haga, H. Three-dimensional morphogenesis of MDCK cells induced by cellular contractile forces on a viscous substrate. *Sci. Rep.* **2015**, *5*, 14208. [CrossRef]
7. Yu, Y.S.; Ahn, C.B.; Son, K.H.; Lee, J.W. Motility Improvement of Biomimetic Trachea Scaffold via Hybrid 3D-Bioprinting Technology. *Polymers* **2021**, *13*, 971. [CrossRef]
8. Chung, I.-M.; Enemchukwu, N.O.; Khaja, S.D.; Murthy, N.; Mantalaris, A.; García, A.J. Bioadhesive hydrogel microenvironments to modulate epithelial morphogenesis. *Biomaterials* **2008**, *29*, 2637–2645. [CrossRef]

9. Wakuda, Y.; Nishimoto, S.; Suye, S.; Fujita, S. Native collagen hydrogel nanofibres with anisotropic structure using core-shell electrospinning. *Sci. Rep.* **2018**, *8*, 6248. [CrossRef] [PubMed]
10. Fujita, S.; Wakuda, Y.; Matsumura, M.; Suye, S. Geometrically customizable alginate hydrogel nanofibers for cell culture platforms. *J. Mater. Chem. B* **2019**, *7*, 6556–6563. [CrossRef]
11. Huang, W.-Y.; Hibino, T.; Suye, S.; Fujita, S. Electrospun collagen core/poly-L-lactic acid shell nanofibers for prolonged release of hydrophilic drug. *RSC Adv.* **2021**, *11*, 5703–5711. [CrossRef]
12. Yang, L.; Shi, J.; Zhou, X.; Cao, S. Hierarchically organization of biomineralized alginate beads for dual stimuli-responsive drug delivery. *Int. J. Biol. Macromol.* **2015**, *73*, 1–8. [CrossRef]
13. Li, H.; Jiang, F.; Ye, S.; Wu, Y.; Zhu, K.; Wang, D. Bioactive apatite incorporated alginate microspheres with sustained drug-delivery for bone regeneration application. *Mater. Sci. Eng. C* **2016**, *62*, 779–786. [CrossRef]
14. Tønnesen, H.H.; Karlsen, J. Alginate in drug delivery systems. *Drug Dev. Ind. Pharm.* **2002**, *28*, 621–630. [CrossRef] [PubMed]
15. Liu, X.-W.; Zhu, S.; Wu, S.-R.; Wang, P.; Han, G.-Z. Response behavior of ion-sensitive hydrogel based on crown ether. *Colloids Surf. A Physicochem. Eng. Asp.* **2013**, *417*, 140–145. [CrossRef]
16. Hu, T.; Lo, A.C.Y. Collagen–Alginate Composite Hydrogel: Application in Tissue Engineering and Biomedical Sciences. *Polymers* **2021**, *13*, 1852. [CrossRef] [PubMed]
17. Chueh, B.; Zheng, Y.; Torisawa, Y.; Hsiao, A.Y.; Ge, C.; Hsiong, S.; Huebsch, N.; Franceschi, R.; Mooney, D.J.; Takayama, S. Patterning alginate hydrogels using light-directed release of caged calcium in a microfluidic device. *Biomed. Microdevices* **2010**, *12*, 145–151. [CrossRef] [PubMed]
18. Rebia, R.A.; Rozet, S.; Tamada, Y.; Tanaka, T. Biodegradable PHBH/PVA blend nanofibers: Fabrication, characterization, in vitro degradation, and in vitro biocompatibility. *Polym. Degrad. Stab.* **2018**, *154*, 124–136. [CrossRef]
19. Qin, Q.; Takarada, W.; Kikutani, T. Fiber Structure Development of PHBH through Stress-Induced Crystallization in High-Speed Melt Spinning Process. *J. Fiber Sci. Technol.* **2017**, *73*, 49–60. [CrossRef]
20. Rebia, R.A.; Shizukuishi, K.; Tanaka, T. Characteristic changes in PHBH isothermal crystallization monofilaments by the effect of heat treatment and dip-coating in various solvents. *Eur. Polym. J.* **2020**, *134*, 109808. [CrossRef]
21. Nomura, S.; Kawai, H.; Kimura, I.; Kagiyama, M. General description of orientation factors in terms of expansion of orientation distribution function in a series of spherical harmonics. *J. Polym. Sci. Part A 2 Polym. Phys.* **1970**, *8*, 383–400. [CrossRef]
22. Batnyam, O.; Suye, S.I.; Fujita, S. Direct cryopreservation of adherent cells on an elastic nanofiber sheet featuring a low glass-transition temperature. *RSC Adv.* **2017**, *7*, 51264–51271. [CrossRef]
23. Hoang, V.T.; Trinh, C.T.; Nguyen, C.H.; Chansomphou, V.; Chansomphou, V.; Tran, T.T.T. Overview of epidermoid cyst. *Eur. J. Radiol. Open* **2019**, *6*, 291–301. [CrossRef]
24. Reissis, D.; Pfaff, M.J.; Patel, A.; Steinbacher, D.M. Craniofacial dermoid cysts: Histological analysis and inter-site comparison. *Yale J. Biol. Med.* **2014**, *87*, 349–357. [PubMed]
25. Ishii, N.; Fukazawa, E.; Aoki, T.; Kishi, K. Combined extracranial and intracranial approach for resection of dermoid cyst of the sphenoid bone with a cutaneous sinus tract across the frontal branch of the facial nerve. *Arch. Craniofacial Surg.* **2019**, *20*, 116–120. [CrossRef] [PubMed]
26. Azab, W.; Salaheddin, W.; Alsheikh, T.; Nasim, K.; Nasr, M. Colloid cysts posterior and anterior to the foramen of Monro: Anatomical features and implications for endoscopic excision. *Surg. Neurol. Int.* **2014**, *5*, 124. [CrossRef] [PubMed]
27. Pedrizzetti, G.; La Canna, G.; Alfieri, O.; Tonti, G. The vortex—an early predictor of cardiovascular outcome? *Nat. Rev. Cardiol.* **2014**, *11*, 545–553. [CrossRef] [PubMed]
28. Brown, B.N.; Badylak, S.F. Extracellular Matrix as an Inductive Scaffold for Functional Tissue Reconstruction. In *Translating Regenerative Medicine to the Clinic*; Elsevier: Amsterdam, The Netherlands, 2016; pp. 11–29. ISBN 9780128005521.
29. Zeug, A.; Stawarski, M.; Bieganska, K.; Korotchenko, S.; Wlodarczyk, J.; Dityatev, A.; Ponimaskin, E. Current microscopic methods for the neural ECM analysis. In *Progress in Brain Research*; Elsevier B.V.: Amsterdam, The Netherlands, 2014; Volume 214, pp. 287–312. ISBN 9780444634863.
30. Edmondson, R.; Broglie, J.J.; Adcock, A.F.; Yang, L. Three-dimensional cell culture systems and their applications in drug discovery and cell-based biosensors. *Assay Drug Dev. Technol.* **2014**, *12*, 207–218. [CrossRef] [PubMed]
31. Lawton, M.T.; Quiñones-Hinojosa, A.; Chang, E.F.; Yu, T. Thrombotic Intracranial Aneurysms: Classification Scheme and Management Strategies in 68 Patients. *Neurosurgery* **2005**, *56*, 441–454. [CrossRef]
32. Iihara, K.; Murao, K.; Sakai, N.; Soeda, A.; Ishibashi-Ueda, H.; Yutani, C.; Yamada, N.; Nagata, I. Continued growth of and increased symptoms from a thrombosed giant aneurysm of the vertebral artery after complete endovascular occlusion and trapping: The role of vasa vasorum. *J. Neurosurg.* **2003**, *98*, 407–413. [CrossRef]

polymers

MDPI

Article

Electrospun Fibers of Polybutylene Succinate/Graphene Oxide Composite for Syringe-Push Protein Absorption Membrane

Nuankanya Sathirapongsasuti [1,2], Anuchan Panaksri [3], Sani Boonyagul [3], Somchai Chutipongtanate [4] and Nuttapol Tanadchangsaeng [3,*]

1   Section of Translational Medicine, Faculty of Medicine Ramathibodi Hospital, Mahidol University, 270 Rama VI Rd., Thung Phaya Thai, Ratchathewi, Bangkok 10400, Thailand; nuankanya.sat@mahidol.edu
2   Research Network of NANOTEC—MU Ramathibodi on Nanomedicine, Bangkok 10400, Thailand
3   College of Biomedical Engineering, Rangsit University, 52/347 Phahonyothin Road, Lak-Hok 12000, Pathumthani, Thailand; anuchan.p59@rsu.ac.th (A.P.); sani@rsu.ac.th (S.B.)
4   Department of Pediatrics, Faculty of Medicine Ramathibodi Hospital, Mahidol University, 270 Rama VI Rd., Thung Phaya Thai, Ratchathewi, Bangkok 10400, Thailand; schuti.rama@gmail.com
*   Correspondence: nuttapol.t@rsu.ac.th; Tel.: +66-(0)2-997-2200 (ext. 1428); Fax: +66-(0)2-997-2200 (ext. 1408)

Citation: Sathirapongsasuti, N.; Panaksri, A.; Boonyagul, S.; Chutipongtanate, S.; Tanadchangsaeng, N. Electrospun Fibers of Polybutylene Succinate/Graphene Oxide Composite for Syringe-Push Protein Absorption Membrane. *Polymers* **2021**, *13*, 2042. https://doi.org/10.3390/polym13132042

Academic Editors: Ick-Soo Kim, Motahira Hashmi and Sana Ullah

Received: 26 May 2021
Accepted: 16 June 2021
Published: 22 June 2021

Publisher's Note: MDPI stays neutral with regard to jurisdictional claims in published maps and institutional affiliations.

Abstract: The adsorption of proteins on membranes has been used for simple, low-cost, and minimal sample handling of large volume, low protein abundance liquid samples. Syringe-push membrane absorption (SPMA) is an innovative way to process bio-fluid samples by combining a medical syringe and protein-absorbable membrane, which makes SPMA a simple, rapid protein and proteomic analysis method. However, the membrane used for SPMA is only limited to commercially available protein-absorbable membrane options. To raise the method's efficiency, higher protein binding capacity with a lower back pressure membrane is needed. In this research, we fabricated electrospun polybutylene succinate (PBS) membrane and compared it to electrospun polyvinylidene fluoride (PVDF). Rolling electrospinning (RE) and non-rolling electrospinning (NRE) were employed to synthesize polymer fibers, resulting in the different characteristics of mechanical and morphological properties. Adding graphene oxide (GO) composite does not affect their mechanical properties; however, electrospun PBS membrane can be applied as a filter membrane and has a higher pore area than electrospun PVDF membrane. Albumin solution filtration was performed using all the electrospun filter membranes by the SPMA technique to measure the protein capture efficiency and staining of the protein on the membranes, and these membranes were compared to the commercial filter membranes—PVDF, nitrocellulose, and Whatman no. 1. A combination of rolling electrospinning with graphene oxide composite and PBS resulted in two times more captured protein when compared to commercial membrane filtration and more than sixfold protein binding than non-composite polymer. The protein staining results further confirmed the enhancement of the protein binding property, showing more intense stained color in compositing polymer with GO.

Keywords: polybutylene succinate; filter membrane; electrospun fiber; graphene oxide; protein adsorption

1. Introduction

With the advances in protein and proteomics analysis technologies, the determination of protein in various bio-fluids has been extensively studied, especially in medical research, in which proteomic profiles or specific protein abundance were reviewed in many diseases [1–5]. In the environmental field, the determination of proteins in water samples could be an indicator of water quality and environmental risk factors [6,7]. Collecting samples from bio-fluids is, therefore, crucial for practical analysis, exclusively for extensive volume sampling. Syringe-push membrane absorption (SPMA) is an innovative way to reduce these problems. SPMA is a method for filtration of substances in liquid samples by filtration through a membrane which is pressurized by pressing the syringe. In this way, the suspended matter in the liquid is fixed on the membrane. Maintenance of the membrane of the sample reduces the sample volume stored for further examination [8]. However, the membrane must effectively capture the analytes or have a small pore size to filter the desired substance. Protein filtration currently uses a PVDF or nitrocellulose membrane to trap proteins [9]. However, the protein absorbance capacity is still low when compared to the original protein content in the fluid samples and the membrane easily tears when pressure is applied.

Therefore, this research aims to develop a membrane with the SPMA technique to have higher protein storage efficiency. The membrane was developed from polybutylene succinate (PBS) polymer, a polymer synthesized from natural materials. PBS is a polymer suitable for packaging, film, and textile applications, biodegrading in the environment [10]. PBS has high strength and flexibility, which is suitable for the pressure applied from the SPMA technique. From the previous study [11], an electrospinning method can be employed to produce a PBS fiber membrane that can be applied in filtration. Electrospinning is a versatile method for fabricating polymer micro- and nano-fibers as components in wound dressings, protective clothing, filtration, and reinforcement fibers in composites [12]. The membranes obtained by this technique increase the surface area and provide micrometer porosity [13]. Their high surface areas are the significant factor dictating their potential in filter membrane applications, which is useful for liquid filtration [14]. In addition, the electrospun PBS fiber membrane has also incorporated graphene oxide into the fabrication. Graphene oxide (GO) is a product obtained by oxidizing graphene. Data on studies of the interactions between graphene oxide and proteins are of interest. Zhang's research [15] shows that graphene oxide can immobilize proteins, signaling that they are due to the chemical function of the two compounds. Graphene contains carboxyl groups, while proteins have amine groups, forming several bonds between chemical functions. This indicates that GO can bond with protein chemical groups, which may be useful for helping to trap proteins.

In this research, we fabricated electrospun PBS and a PBS/GO composite membrane, comparing it to electrospun PVDF, a material currently used with the SPMA technique, and PVDF/GO with the different fabrication methods. The two types of collector methods—rolling electrospinning (RE) and non-rolling electrospinning (NRE)—were employed to synthesize polymer fibers, resulting in different characteristics of mechanical and morphological properties. Moreover, all the electrospun filter membranes were performed during the filtration of the albumin solution by the SPMA technique to measure the protein capture efficiency as well as staining of the protein on the filter membrane, and were also compared with commercial filter membranes of PVDF, nitrocellulose and Whatman no. 1 paper.

2. Materials and Methods

2.1. Materials

Polybutylene succinate (PBS) (FD92) was supplied by PTT MCC Biochem Ltd (Bangkok, Thailand). Poly(vinylidene fluoride) (PVDF) was purchased from Sigma-Aldrich (batch no. 427144, St. Louis, MO, USA), with an average molecular weight of 275 kg/mol, according to the manufacturer. Graphene oxide powder, 15–20 sheets, 4–10% edge-oxidized, was purchased from Sigma-Aldrich (St. Louis, MO, USA). Dimethyl sulfoxide (DMSO) was purchased from Sigma-Aldrich (St. Louis, MO, USA). Acetone, methanol and chloroform were purchased from RCI Labscan (Bangkok, Thailand). Brilliant Blue Coomassie G-250 was purchased from ACROS Organics™. Hybridization Nitrocellulose Filter membrane (filter type—0.45 µm HATF) was purchased from Merck Millipore Ltd., Cork, Ireland. Transfer PVDF membrane (Immobion-P Transfer membrane, pore size 0.45 µm) was purchased from EMD Millipore Corporation, Burlington, MA, USA. Whatman no. 1 filter papers were purchased from GE Healthcare Life Sciences, Marlborough, MA, USA. Human albumin was purchased from Sigma-Aldrich, St. Louis, MO, USA.

2.2. Methods

2.2.1. Solution Preparation for Electrospun Fiber Composite Membrane

The 20 wt.% PVDF solution was prepared by dissolving PVDF in a mixture of DMSO and acetone (DMSO/acetone) 7:3 (volume ratio), with heating at 50 °C for 24 h. The 8 wt.% PBS solution was prepared in the same way as the PVDF, but the solvent was changed to chloroform/methanol 7:3 (volume ratio). The prepared polymer solution was divided into two conditions: with and without graphene oxide. A composite polymer solution was added with graphene oxide for 0.1% by weight of the polymer, mixed at a temperature of 60 °C, and then, subsequently underwent 15 min sonication.

2.2.2. Fabrication of Electrospun Fiber Composite Membrane by Electrospinning

The prepared polymer solution was fabricated into a filter membrane using a single nozzle electrospinning setup with a flat grounded collector (non-rolling electrospinning, NRE) or rotating drum collector (rolling electrospinning, RE) with the speed set at 500 rpm. A high voltage power supply (ES60P-10 W, Gamma High Voltage, Ormond Beach, FL, USA) was attached to a metal nozzle to create a voltage difference between the metal nozzle and a collector of 25 kV. A charged metal nozzle with 0.57 mm outer diameter was placed 15 cm directly from the collector covered with aluminum foil. A mechanical pump (NE-300 Just Infusion™ Syringe Pump New Era Pump Systems, Inc., Farmingdale, NY, USA) was used to deliver the solution from a plastic syringe to the metal nozzle at a rate of 3 mL/h, with a feeding time of 4 h.

2.2.3. Mechanical Properties Test and Uniformity Measurement of Membranes

A rectangular 1×2 cm^2 membrane sample attached at both ends with 2×3 cm^2 sandpaper was prepared. The mechanical properties of the electrospun and commercial membranes were evaluated using a universal test machine (TestResources, Shakopee, MN, USA) with a 5 kg load cell equipped with tensile grips. The stress–strain tests with a strain rate of 10 mm/min were performed until fracture occurred, and the data were obtained in an average of three measurements. Membrane size uniformity was measured from a standard deviation of membrane mean thickness. The membrane thickness was measured with a precise micrometer, randomly measuring five different points.

2.2.4. Examination of Morphology by Scanning Electron Microscopy

A scanning electron microscope (SEM) (Prisma E, Thermo Scientific, Waltham, MA, USA) was used to characterize the electrospun fibers' morphology and the size of the prepared 2×2 cm^2 samples. Samples of the electrospun membranes were sputter-coated with gold. Images were captured with an accelerating voltage of 5 kV and were recorded at 16000× magnification. Image J software (www.imagej.nih.gov, Accessed on 1 May 2021) was used for measuring fiber diameter and pore size. The average fiber diameter and average pore area with standard deviation (SD) were determined from ten measurements.

2.2.5. Comparison of Membrane Protein Binding Capacity

Three membranes of each condition were cut in a circular shape, 13 mm in diameter (n = 3). An amount of 1 mg/mL human albumin solution was prepared in deionized (DI) water. Then, 5 mL of prepared albumin solution was pushed through each filter membrane by the SPMA technique. The membrane adsorption capacity was relatively calculated from flowed through protein. An amount of 10 μL of each membrane flow-through fluid was used for protein measurement using a Bradford reagent kit (Bio-Rad), according to the manufacturer's protocol. The flow-through protein concentrations of each filter membrane are shown in Figure S1. The albumin proteins attached to the electrospun filter (% protein adsorption) were calculated by Equation (1).

$$\% \textbf{\textit{Protein adsorption}} = \left(\frac{\textbf{1} - \textbf{\textit{x}}}{\textbf{1}} \right) \times \textbf{100} \qquad (1)$$

where 1 is 100% albumin in the testing solution, and x is the amount of albumin that passed through the filter.

2.2.6. Staining of Protein on the Filtered Specimens

Protein staining was prepared using a Neuhoff dye containing 2% H_3PO_4, 10% $(NH_4)_2SO_4$, 20% methanol, and 0.1% Coomassie G-250, respectively. A total of 200 μL of dye solution was pipetted on the filter membrane that had already undergone the SPMA technique. After 30 min, the dye reagent was rinsed off with DI water. After that, the filter specimens were dried in the air, and then, photographs were taken for qualitative investigation of the protein on each filter membrane.

2.2.7. Statistical Analysis

Data were analyzed using the SPSS Statistics software version 25.0 (IBM, Armonk, NY, USA) and Microsoft Excel 2019 (Microsoft, Redmond, WA, USA). A paired t-test was used for the comparison of two sets of data. The differences between means at the 95% confidence level ($p < 0.05$) were considered statistically significant.

3. Results and Discussion

3.1. Mechanical Properties and Membrane Thickness Uniformity

Table 1 shows the mechanical properties and the thickness of the various fabricated membrane conditions, depending on the fiber collection method and polymer types. The PBS and PVDF fibers via electrospinning were processed by high voltage on the needle and zero voltage on both collectors (flat grounded collector (non-rolling electrospinning, NRE) and rotating drum collector (rolling electrospinning, RE)). It can be noted that the PBS/solvent droplet can be pushed by high voltage on the needle to the collector; however, the droplet had no ability to form fibers on the flat collector. These may result from a solution jet that rapidly elongates the solution into micro- or nano-size until the solvent is completely evaporated, before the fiber drops on the collector. However, for rotating drum collectors, both PBS and PVDF could form fibers with the rolling electrospinning technique.

Table 1. Sample thickness and mechanical properties of each filter membrane used in this study.

Sample	Thickness (μm)	Young's Modulus (MPa)	Max. Tensile Strength (MPa)	Elongation at Break (%)
PBS (RE)	30.8 ± 7.9	193 ± 3.0	24 ± 3.6	58 ± 6.5
PBS + GO 0.1% (RE)	45.2 ± 6.9	170 ± 2.0	31 ± 4.6	50 ± 2.6
PVDF (NRE)	116.8 ± 47.3	143 ± 5.6	18 ± 3.6	61 ± 7.2
PVDF + GO 0.1% (NRE)	87.6 ± 47.3	101 ± 2.6	24 ± 2.0	53 ± 5.2
PVDF + GO 0.1% (RE)	82.4 ± 3.2	181 ± 2.0	32 ± 4.0	57 ± 5.2
Nitrocellulose (commercial)	140.4 ± 1.0	4.8 ± 0.6	0.18 ± 0.1	9 ± 2.6
PVDF (commercial)	120.2 ± 1.0	5.3 ± 1.2	0.12 ± 0.1	9 ± 3.5

Note: RE is rolling electrospinning, NRE is non-rolling electrospinning.

As shown in Table 1, the electrospun fiber membrane fabricated by NRE exhibits non-uniform thickness, while the RE membranes have uniformity. The uneven thickness of the NRE membrane indicates that the pattern in the membrane's fabrication had low consistency due to the inability to direct the injected fibers to the plate collector. However, for rolling electrospinning (RE), the injected fibers could be controlled, to some extent, due to the presence of a rotating drum, which set the fiber direction to uniformity [16]. The electrospun membranes of all conditions have a substantially higher value of each mechanical property of Young's modulus, maximum tensile strength, and elongation at break than commercial PVDF and nitrocellulose membranes, indicating more suitable strength for protein filtration. For all electrospun membranes, the addition of graphene oxide in PBS and PVDF polymers did not affect to the specimen's mechanical behaviors, which were consistent with the study of PBS composited with graphene [17]. In the comparison of non-rolling electrospinning and rolling electrospinning, it was shown that the membranes from RE had a higher Young's modulus than the NRE membranes. When applying pressure, the SPMA technique was typically used to filter proteins pressurized from the syringe through a membrane, where the membrane must be strong and flexible to support the force. Both polymers respond well to the above properties. It is indicated that both electrospun PBS and PVDF, with or without GO composite membranes, could tolerate the pushed pressure.

3.2. Morphology of the Fabricated Membranes

The images of the electrospun fibers of the membrane in each condition through a scanning electron microscope (SEM) are shown in Figure 1.

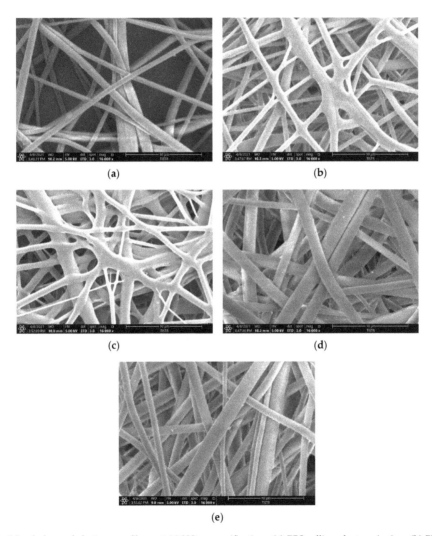

Figure 1. Morphology of electrospun fibers at 16,000× magnification. (**a**) PBS rolling electrospinning, (**b**) PBS + GO 0.1% rolling electrospinning, (**c**) PVDF non-rolling electrospinning, (**d**) PVDF + GO 0.1% non-rolling electrospinning, and (**e**) PVDF + GO 0.1% rolling electrospinning.

Figure 1 shows the characteristics of the fibers in 16,000× SEM images. The nature of the fibers formed by different methods with and without GO composite indicates the fiber overlap patterns and the size of the fiber gaps. However, we can detect the GO fragments trapped on the expanding fibers in the graphene oxide-containing membranes. The morphologies of the nanofibers obtained through RE and NRE of PBS and PVDF are the same (all fine, smooth, and cylindrical fibers), but the diameter changes. As listed in Table 2, the average diameters of the PBS rolling electrospinning, PBS + GO 0.1% rolling electrospinning, PVDF non-rolling electrospinning, PVDF + GO 0.1% non-rolling electrospinning, and PVDF + GO 0.1% rolling electrospinning were 1.0, 0.87, 0.9, 1.17 and 1.05 μm, respectively, indicating that each condition had a similar average diameter, with PBS + GO 0.1% membranes having the smallest fibers. However, from the pore area measured with ImageJ software, PBS membranes were slightly smaller than PVDF membranes. These results indicate that the electrospun PBS membrane has a higher ability to be employed as a filter membrane than the electrospun PVDF membrane.

Table 2. Fiber diameter and membrane area from SEM images.

Sample	Average Diameter of Fibers (μm)	Average Pore Area (μm^2)
PBS (RE)	1.00 ± 0.27	27.01 ± 4.48
PBS + GO 0.1% (RE)	0.87 ± 0.16	23.32 ± 2.98
PVDF (NRE)	0.90 ± 0.29	36.26 ± 5.86
PVDF + GO 0.1% (NRE)	1.17 ± 0.22	37.48 ± 6.25
PVDF + GO 0.1% (RE)	1.05 ± 0.22	33.29 ± 8.33

Membrane morphology substantially confirmed the hypothesis of the previously discussed mechanical properties. The morphology also suggests that variations can occur with rolling electrospinning, possibly depending on the polymer used in fabrication. Morphological results suggest that the fiber size and the porosity for each condition were similar to the previous research, with a fiber size of approximately 1 μm [18]. Commercial membranes (nitrocellulose membranes and PVDF membranes) have a smaller pore size than electrospinning membranes of PBS or PBS+GO. However, the electrospun membranes have an advantage over the contact surface area [19], which could increase protein adsorption ability. The morphology of PBS + GO 0.1% (RE) combined branched-flat and straight-round fibers, while that of PVDF + GO 0.1% (RE) contained only straight-flat and straight-round fibers. These phenomena may result in a higher area that can adsorb the small molecules of protein, giving rise to an increase in protein binding capacity of PBS + GO 0.1% (RE).

3.3. The Protein Binding Capacity of the Fabricated Filter Membranes and Commercially Available Filter Membranes

The protein binding capacity of the fabricated filter membranes and commercially available filter membranes were compared based on a relative calculation from remnant protein in the flow-through fluid, as shown in Figure 2. The protein binding capacity was calculated using Equation (1) (details in Materials and Methods). Five different fabricated conditions and three commercial filter membranes were used for comparison of protein binding capacity. The experiments were performed in triplicate for each membrane. The results showed that Whatman no. 1 filter paper had the lowest protein binding capacity, followed by pure PVDF (NRE), and pure PBS (RE) with protein binding capacity of 2.76%, 5.48%, and 6.47%, respectively. The nitrocellulose showed the highest protein binding capacity among commercially available filter membranes, with a binding capacity of 21.32%. Using the same polymer, the pure PVDF filter membrane developed in this study showed less protein absorbance when compared with the commercial one, 5.48% and 14.24%; this might be because of the pore size, 0.90 ± 0.29 μm and 0.45 μm, and the random pattern of the electrospinning.

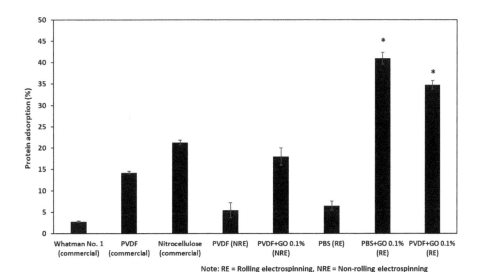

Figure 2. The protein adsorption capacity of five fabricated membranes and three commercial membranes used in this study. Note: * Significant differences of the two samples compared to the others.

Interestingly, the addition of GO could improve the protein binding capacity dramatically to 3.29-fold when using non-rolling electrospinning and even higher to 6.5-fold with the rolling electrospinning. The results of PBS also showed a similar trend; adding GO coupled with the rolling electrospinning process could increase protein adsorption to 6.32-fold when compared with PBS without GO using the same manufacturing process. Taking all the results together, the PBS with GO 0.1% rolling electrospinning showed the highest protein capacity, about 2-fold higher than commercial nitrocellulose membrane and 2.88-fold higher when compared to the commercial PVDF membrane.

3.4. Qualitative Investigation of Absorbed Protein on the Membrane by Protein Staining

Protein staining with Neuhoff dye was applied to determine the distribution and absorbed protein on each filtered membrane. The blue color on each sample indicates the protein's attachment on the membrane. The results of dyeing are shown in Figure 3. DI water filtration was used for baseline staining of the materials (blank control).

The results showed that PBS + GO 0.1% rolling electrospinning, PVDF + GO 0.1% rolling electrospinning, PVDF + GO 0.1% non-rolling electrospinning, and nitrocellulose had deep blue staining after albumin adsorption as compared to fainter blue staining after DI water adsorption. Interestingly, the graphene oxide-based membranes showed more intense dye staining than that of non-graphene-added samples. Commercial membranes (PVDF and nitrocellulose) exhibited lighter blue staining than those containing graphene oxide. Moreover, the fabricated procedure, rolling electrospinning, and non-rolling electrospinning resulted in different adsorption capacities. The Whatman no. 1 filter paper, as the negative control, presented very light blue staining in both protein and blank control filtering, indicating very low protein adsorption ability. The PBS + GO 0.1% still has the highest value of color intensity, compared to the other intensities of the protein staining data, as shown in Figure S2. These quantitative intensity results are consistent with the data of the protein adsorption experiment. These results indicate that the membrane morphology also supported the reasoning in terms of fabrication patterns and three-dimensional structure that increase surface area and aid in protein capture [20].

Sample	5 mL albumin solution	5 mL DI water
PBS (RE)		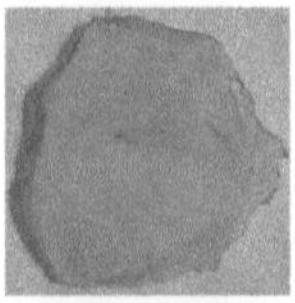
PBS + GO 0.1% (RE)		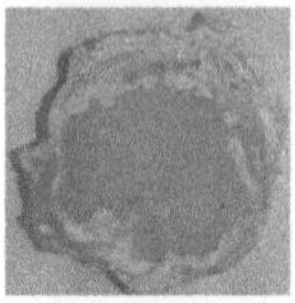
PVDF (NRE)	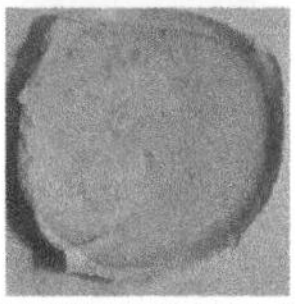	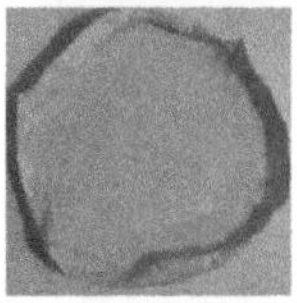
PVDF + GO 0.1% (NRE)	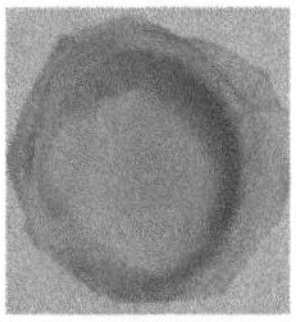	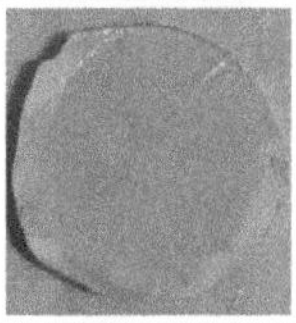
PVDF + GO 0.1% (RE)	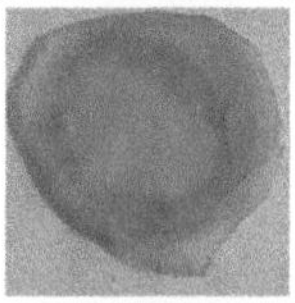	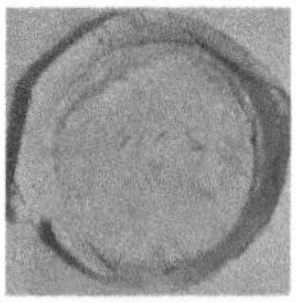
PVDF membrane (commercial)	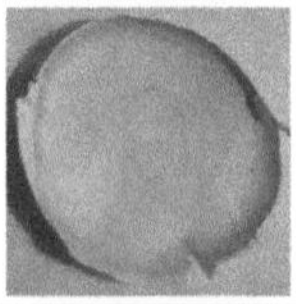	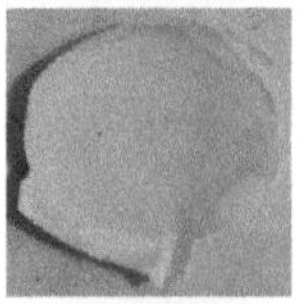
Nitrocellulose membrane (commercial)	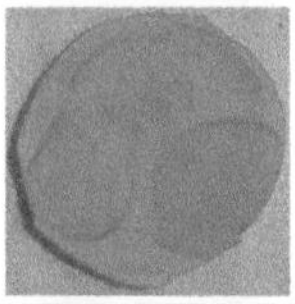	
Whatman no. 1 (commercial)	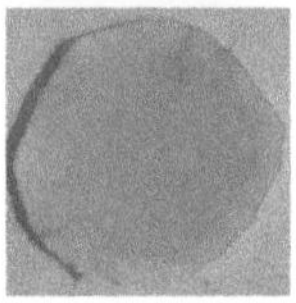	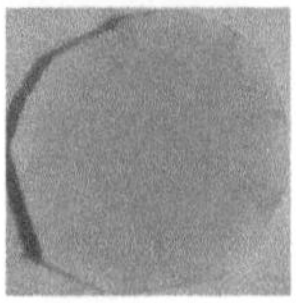

Figure 3. Brilliant blue Coomassie G-250 staining of absorbed proteins on membranes.

4. Conclusions

Syringe-push membrane absorption (SPMA) is a rapid and straightforward method of bio-fluid preparation for protein and proteomic study. However, the commercially available membranes still have some drawbacks: low protein adsorption capacity and tearing when pressure is applied. This research, employing PBS composite with GO and fabricating electrospun membrane on the rolling electrospinning, resulted in the best performance and could solve previous SPMA issues. Further scaled-up production, protein recovery optimization, and recovered protein yield evaluation would be crucial steps toward practical implementation. To the best of our knowledge, this research is the first experiment that reports the utilization of the electrospun bioplastic as a protein adsorption membrane for further biomedical applications.

Supplementary Materials: The following are available online at https://www.mdpi.com/article/10.3390/polym13132042/s1, Figure S1: Calibration curve of Protein vs. OD. Figure S2: Color intensities of the staining membranes.

Author Contributions: Conceptualization, N.S., S.C. and N.T.; Formal analysis, N.S., A.P. and N.T.; Funding acquisition, N.S. and N.T.; Investigation, A.P., S.B. and N.T.; Methodology, N.S. and N.T.; Project administration, N.S. and N.T.; Resources, S.B. and N.T.; Visualization, N.S., A.P. and N.T.; Writing—original draft, N.S., A.P. and N.T.; Writing—review and editing, N.S. and N.T. All authors have read and agreed to the published version of the manuscript.

Funding: This research project was supported by the Research Network NANOTEC program (RNN-Ramathibodi hospital) and the Faculty of Medicine Ramathibodi Hospital Mahidol University (grant number CF_62006).

Institutional Review Board Statement: Not applicable.

Informed Consent Statement: Not applicable.

Data Availability Statement: The data presented in this study are available on request from the corresponding author.

Acknowledgments: We especially would like to thank Wararat Chiangjong and Kasetsin Khonchom, Faculty of Medicine Ramathibodi Hospital, for preparing the commercial filter membranes and human albumin.

Conflicts of Interest: The authors declare no conflict of interest.

References

1. Senturk, A.; Sahin, A.T.; Armutlu, A.; Kiremit, M.C.; Acar, O.; Erdem, S.; Bagbudar, S.; Esen, T.; Tuncbag, N.; Ozlu, N. Quantitative proteomics identifies secreted diagnostic biomarkers as well as tumor-dependent prognostic targets for clear cell Renal Cell Carcinoma. *Mol. Cancer Res. MCR* **2021**. [CrossRef] [PubMed]
2. Swensen, A.C.; He, J.; Fang, A.C.; Ye, Y.; Nicora, C.D.; Shi, T.; Liu, A.Y.; Sigdel, T.K.; Sarwal, M.M.; Qian, W.J. A Comprehensive Urine Proteome Database Generated From Patients With Various Renal Conditions and Prostate Cancer. *Front. Med.* **2021**, *8*, 548212. [CrossRef]
3. Vanarsa, K.; Enan, S.; Patel, P.; Strachan, B.; Sam Titus, A.; Dennis, A.; Lotan, Y.; Mohan, C. Urine protein biomarkers of bladder cancer arising from 16-plex antibody-based screens. *Oncotarget* **2021**, *12*, 783–790. [CrossRef]
4. Verbeke, F.; Siwy, J.; Van Biesen, W.; Mischak, H.; Pletinck, A.; Schepers, E.; Neirynck, N.; Magalhães, P.; Pejchinovski, M.; Pontillo, C.; et al. The urinary proteomics classifier chronic kidney disease 273 predicts cardiovascular outcome in patients with chronic kidney disease. *Nephrol. Dial. Transplant. Off. Publ. Eur. Dial. Transpl. Assoc. Eur. Ren. Assoc.* **2021**, *36*, 811–818. [CrossRef] [PubMed]
5. Wang, Y.; Zhang, J.; Song, W.; Tian, X.; Liu, Y.; Wang, Y.; Ma, J.; Wang, C.; Yan, G. A proteomic analysis of urine biomarkers in autism spectrum disorder. *J. Proteom.* **2021**, *242*, 104259. [CrossRef] [PubMed]
6. Le, C.; Kunacheva, C.; Stuckey, D.C. "Protein" Measurement in Biological Wastewater Treatment Systems: A Critical Evaluation. *Environ. Sci. Technol.* **2016**, *50*, 3074–3081. [CrossRef]
7. Westgate, P.J.; Park, C. Evaluation of Proteins and Organic Nitrogen in Wastewater Treatment Effluents. *Environ. Sci. Technol.* **2010**, *44*, 5352–5357. [CrossRef]
8. Chutipongtanate, S.; Changtong, C.; Weeraphan, C.; Hongeng, S.; Srisomsap, C.; Svasti, J. Syringe-push membrane absorption as a simple rapid method of urine preparation for clinical proteomics. *Clin. Proteom.* **2015**, *12*, 15. [CrossRef] [PubMed]
9. Jia, L.; Liu, X.; Liu, L.; Li, M.; Gao, Y. Urimem, a membrane that can store urinary proteins simply and economically, makes the large-scale storage of clinical samples possible. *Sci. China Life Sci.* **2014**, *57*, 336–339. [CrossRef] [PubMed]
10. Sharma, R.; Jafari, S.M.; Sharma, S. Antimicrobial bio-nanocomposites and their potential applications in food packaging. *Food Control.* **2020**, *112*, 107086. [CrossRef]

11. Liu, Y.; He, J.H.; Yu, J.Y. Preparation and morphology of poly(butylene succinate) nanofibers via electrospinning. *Fibres Text. East. Eur.* **2007**, *15*, 30–33.
12. Hu, X.; Liu, S.; Zhou, G.; Huang, Y.; Xie, Z.; Jing, X. Electrospinning of polymeric nanofibers for drug delivery applications. *J. Control. Release* **2014**, *185*, 12–21. [CrossRef] [PubMed]
13. Rahmati, M.; Mills, D.K.; Urbanska, A.M.; Saeb, M.R.; Venugopal, J.R.; Ramakrishna, S.; Mozafari, M. Electrospinning for tissue engineering applications. *Prog. Mater. Sci.* **2021**, *117*, 100721. [CrossRef]
14. Ko, F.K.; Wan, Y. *Introduction to Nanofiber Materials*; Cambridge University Press: Cambridge, UK, 2014.
15. Zhang, Y.; Wu, C.; Guo, S.; Zhang, J. Interactions of graphene and graphene oxide with proteins and peptides. *Nanotechnol. Rev.* **2013**, *2*, 27–45. [CrossRef]
16. Xue, J.; Wu, T.; Dai, Y.; Xia, Y. Electrospinning and Electrospun Nanofibers: Methods, Materials, and Applications. *Chem. Rev.* **2019**, *119*, 5298–5415. [CrossRef]
17. Platnieks, O.; Gaidukovs, S.; Neibolts, N.; Barkane, A.; Gaidukova, G.; Thakur, V.K. Poly(butylene succinate) and graphene nanoplatelet–based sustainable functional nanocomposite materials: Structure-properties relationship. *Mater. Today Chem.* **2020**, *18*, 100351. [CrossRef]
18. Huang, L.; Arena, J.T.; McCutcheon, J.R. Surface modified PVDF nanofiber supported thin film composite membranes for forward osmosis. *J. Membr. Sci.* **2016**, *499*, 352–360. [CrossRef]
19. Grafe, T.; Graham, K. Polymeric Nanofibers and Nanofiber Webs: A New Class of Nonwovens. *Int. Nonwovens J.* **2003**, *os-12*, 1558925003os-1551200113. [CrossRef]
20. Nematollahzadeh, A.; Shojaei, A.; Abdekhodaie, M.J.; Sellergren, B. Molecularly imprinted polydopamine nano-layer on the pore surface of porous particles for protein capture in HPLC column. *J. Colloid Interface Sci.* **2013**, *404*, 117–126. [CrossRef] [PubMed]

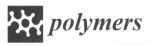 *polymers*

Article

Green Synthesis and Incorporation of Sericin Silver Nanoclusters into Electrospun Ultrafine Cellulose Acetate Fibers for Anti-Bacterial Applications

Mujahid Mehdi [1,2], Huihui Qiu [1], Bing Dai [1], Raja Fahad Qureshi [2,3], Sadam Hussain [1,2], Muhammad Yousif [1], Peng Gao [1,*] and Zeeshan Khatri [2,*]

[1] National Engineering Laboratory for Modern Silk, College of Textile and Clothing Engineering, Soochow University, Suzhou 215123, China; mujahid11te83@gmail.com (M.M.); kitty.huihui@aliyun.com (H.Q.); db@szdfhj.cn (B.D.); sadam11te75@gmail.com (S.H.); muhammadyousif.te72@gmail.com (M.Y.)

[2] Center of Excellence in Nanotechnology and Materials, Mehran University of Engineering and Technology, Jamshoro 76060, Pakistan; raja.ashraf@faculty.muet.edu.pk

[3] Department of Textile Engineering, Mehran University of Engineering and Technology, Jamshoro 76060, Pakistan

* Correspondence: pgao116@126.com (P.G.); zeeshan.khatri@faculty.muet.edu.pk (Z.K.); Tel.: +86-15862320382 (P.G.); +92-(0)22-2772250 (Z.K.)

Citation: Mehdi, M.; Qiu, H.; Dai, B.; Qureshi, R.F.; Hussain, S.; Yousif, M.; Gao, P.; Khatri, Z. Green Synthesis and Incorporation of Sericin Silver Nanoclusters into Electrospun Ultrafine Cellulose Acetate Fibers for Anti-Bacterial Applications. *Polymers* **2021**, *13*, 1411. https://doi.org/10.3390/polym13091411

Academic Editor: Andrea Ehrmann

Received: 1 April 2021
Accepted: 25 April 2021
Published: 27 April 2021

Abstract: Fiber based antibacterial materials have gained an enormous attraction for the researchers in these days. In this study, a novel Sericin Encapsulated Silver Nanoclusters (sericin-AgNCs) were synthesized through single pot and green synthesis route. Subsequently these sericin-AgNCs were incorporated into ultrafine electrospun cellulose acetate (CA) fibers for assessing the antibacterial performance. The physicochemical properties of sericin-AgNCs/CA composite fibers were investigated by transmission electron microscopy (TEM), field emission electron microscopy (FE-SEM), Fourier transform infrared spectroscopy (FTIR) and wide X-ray diffraction (XRD). The antibacterial properties of sericin-AgNCs/CA composite fibers against *Escherichia coli* (*E. coli*) and *Staphylococcus aureus* (*S. aureus*) were systematically evaluated. The results showed that sericin-AgNCs incorporated in ultrafine CA fibers have played a vital role for antibacterial activity. An amount of 0.17 mg/mL sericin-AgNCs to CA fibers showed more than 90% results and elevated upto >99.9% with 1.7 mg/mL of sericin-AgNCs against *E. coli*. The study indicated that sericin-AgNCs/CA composite confirms an enhanced antibacterial efficiency, which could be used as a promising antibacterial product.

Keywords: silver nanoclusters; sericin; cellulose acetate; electrospinning; Antibacterial Nanofibers

1. Introduction

Globally, a new and emerging variety of infectious diseases have brought the human health under a big threat. In most of the cases, the bacterial infection not only cause severe sickness accompanied by developing viral influenza (such as SARs and avian flu), but these microorganisms can also be spread into the environment making it rather unsafe [1,2]. For this reason, exploring new antimicrobial agents have gained much attention in past few decades. Generally, the metallic nanoparticles (Ag, Au, Cu, and Pt) and metallic oxides nanoparticles (Ag_2O, CuO, ZnO, TiO_2, Fe_2O_3, and SiO_2) has been reported for antibacterial applications [3]. Amongst these nanoparticles, the silver nanoparticles are notably preferred due to their strong antimicrobial efficiency against microorganisms [4]. It is believed that the bacterial cell can be destroyed upon its the interaction with the Ag ion. The positively charged Ag ions, due to their strong affinity to sulfur proteins, adhere to the cell wall and cytoplasmic membrane [5–7]. This results into the blockage of bacterial respiratory system and demolishing the cell production. Subsequently this leads to the remarkable decline into the growth of further bacteria [8,9].

In a comparison with Ag particles, Ag ultra-small nanoparticles are improved in chemical stability, heat resistance, release of Ag ions, and durability, which will ultimately bring more opportunities for antimicrobial products [10]. Previous investigations revealed that the size [11], shape [12], surface coating [13], and surface charge [14] of silver ultra-small nanoparticles can significantly affect antibacterial efficacy. The novel metal (Au, Ag, Cu, and Pt) nanoclusters (NCs) composed of few to hundreds of atoms exhibited ultra-small sizes (<5 nm) that approach to the fermi wavelength of the conduction electrons. They show discrete energy levels, resulting in obvious fluorescence and molecule-like properties [15,16]. Thus, these NCs possess very unique properties than that of traditional metal nanoparticles and have attracted great attraction in recent years. These NCs have been extensively investigated in biolabeling and bioimaging, medicine, catalysis, and nanoelectronics [17,18]. Therefore, it is anticipated that silver nanoclusters (AgNCs) may also exhibit superior antimicrobial properties. Recently, Jian-Cheng group has reported the antibacterial performance of ultra-small DHLA-AgNCs and the results showed that silver ions of rich NCs could act as efficient nano reservoirs for antibacterial activity [19,20]. Sericin is extracted from silk fiber (*Bombyx mori* silkworm) and present in the outer layer of silk fiber. Sericin contains 18 types of amino acids, such as serine, glycine, lysine, etc. Sericin is a biocompatible and biodegradable natural biopolymer exhibiting antioxidant, moisture absorption, antibacterial and UV resistance properties [21,22].

Recently, safe and advanced non-woven nanofibrous dressings have received tremendous attention in antibacterial applications [23]. It is also stated that nanofibers have remarkable properties such as high interconnected porosity, large surface area to volume ratios [24,25]. To enhance the stability and therapeutic efficacy, Ag nanoparticles has been incorporated into various polymeric nanofibers including cellulose acetate (CA) [26], polyurethane [27], poly (ethylene oxide) [28], poly (vinyl alcohol) [29], and poly (lactic-co-glycolic acid) [30]. While CA contains polysaccharides groups from cellulose materials, it is a commonly used biopolymer and promising material for medical applications. Many conventional metal nanoparticles incorporating into electrospun nanofibers has been prepared. However, there is a considerable need to enhance antibacterial efficiency of electrospun membrane and its biocompatibility. Due to their ultra small size, it is assumed that metal nanoclusters can fulfil the above purpose.

In the present study, sericin mediated silver nanoclusters were prepared and incorporated into ultrafine CA nanofibers in order to assess the antibacterial activity of sericin-AgNCs/CA composite membrane against against *S. aureus* and *E. coli*. It was found that there is significant effect of sericin-AgNCs on the performance of antibacterial activity due to ultra-small size of Ag particles. The physicochemical properties of sericin-AgNCs incorporated into ultrafine CA fibers were examined with transmission electron microscopy (TEM), field emission electron microscopy (FE-SEM), Fourier transform infrared spectroscopy (FTIR) and wide X-ray diffraction (XRD). To the best of our knowledge, sericin-AgNCs were synthesized for the very first time, and incorporated into ultrafine CA fibers for antibacterial application.

2. Experimental

2.1. Materials

Cellulose acetate (CA, 30 KDa), Silver Nitrate ($AgNO_3$, 99.9%), and sericin were purchased from Sigma Aldrich Co., Ltd. (St. Louis, MO, USA). Sodium hydroxide (NaOH, 99%), Acetone (C_3H_6O, 99.5%) and Dimethyl formamide (C_3H_7NO, 99.5%) were purchased from Sinopharm Chemical Reagent Co., Ltd. (Shanghai, China). Ultra-pure water was used during all experiments.

2.2. Synthesis of Sericin-AuNCs

The template of Ag Nanoclusters with sericin was synthesized by using $AgNO_3$ regent. For this, 20 mL of sericin (50 mg/mL) and 20 mL of $AgNO_3$ (10 mM) was prepared separately and both solutions were transferred into an incubator at 37 °C for 10 min.

Afterwards, both solutions were mixed together and shaken gently to get a homogenous solution of the mixture. Then, 2 mL NaOH (1 M) was added in the mixture and kept on stirring for 3 min. The solution was later incubated at 37 °C for 16 h and then placed further for 12 h into refrigerator at −20 °C. Finally, the sample was placed for freeze drying for 2 days to achieve Sericin-AgNCs in powder form.

2.3. Electrospinning

The ultrafine CA fibers were fabricated according to our previous method [31]. Briefly, the polymer solution consists of 17% CA polymer was dissolved in acetone: DMF with ratio of 2:1 respectively. Then, different amounts of synthesized sericin-AgNCs powder (0.17 mg/mL, 0.85 mg/mL and 1.7 mg/mL) was added into the polymeric solution and stirred for 24 h in order to get homogeneous solution. The prepared solution was filled in a 5 mL syringe and placed vertically on micro-injection pump (LSP02-1B, Baoding longer precision pump Co., Ltd., Baoding, China). The feed rate for the solution was adjusted at a constant and controllable rate of 0.5 mL/h. High voltage of 10 kV was applied using power supply (DWP303-1AC, Tianjin Dongwen High Voltage Co., Baoding, China). The distance between needle and collector was set at 15 cm. The fibers were collected on aluminum foil and after collecting fibers, samples were dried in open air for 24 h.

2.4. Antibacterial Activity

The antibacterial performance was analyzed using the common shake flask protocol [32]. This protocol is specifically designed for specimen's treatment with non-releasing antibacterial agents under dynamic contact conditions. *Escherichia coli* (BUU25113) and *Staphylococcus aureus* (B-sub 168) were cultivated in nutrient broth at 37 °C for 18 h and the prepared solution was examined by broth dilution method. The number of viable bacterial cells reached 1×10^9 cfu/mL. During four times serial dilution with 0.03 mol/L PBS, the number of viable bacterial cells is adjusted 3×10^5 cfu/mL to 4×10^3 cfu/mL. After obtaining the desired growth of bacterial cells, 0.05 g sample (CA fibers) was poured into conical flask containing 65 mL of 0.3 mM PBS solution and 5 mL of the prepared solution containing bacterial growth. This mixed up solution was then placed on shaking machine at 37 °C for 18 h. After shaking, the process was further repeated upto four times for serial dilution by mixing 1 mL of solution from the flask and 9 mL of 0.3 mM PBS. Finally 1 mL of the solution of bacterial growth with different concentration was taken and placed onto an agar plate. After 24 h of incubation at 37.8 °C, the number of bacterial colonies formed on the agar plates were counted visually. Moreover, the antibacterial performance was established from obtained results and percentage bacterial reduction was calculated according to following equation.

$$R = \frac{(B - A)}{B} \times 100\% \tag{1}$$

where R is the percentage bacterial reduction, B and A are the number of residual colonies of before and after treated with ultrafine CA fibers.

2.5. Material Analysis

The surface morphology and average diameter of template sericin-AuNCs were analyzed using transmission electron microscope (TEM, H-800 Hitachi Ltd., Tokyo, Japan) operating at 200 kV. The average particle size of sericin Ag NCs was measured from individual particles on the TEM images. Surface morphology of ultrafine electrospun CA fibers and sericin-AgNCs incorporated into ultra-fine CA fibers was measured using Field emission scanning electronic microscopy (FE-SEM, S-4800 Hitachi Ltd., Tokyo, Japan). The average fiber diameter was measured using image J software and graphical presented by counting 50 distinct fibers from FE-SEM images. The SEM samples were sputtered with gold before examination. Chemical composition of sericin-AgNCs incorporated into ultrafine CA fiber was obtained using energy-dispersive X-ray spectroscopy (EDS, S-3000

N Hitachi Ltd., Tokyo, Japan) elemental analysis, showing carbon, oxygen and silver as the main elements. Fourier transform infrared spectroscopy (FTIR) measurement was analyzed with Thermo Fisher Scientific Inc., Waltham, MA, USA. All the FTIR samples 1–2 mg were ground with 0.1 g KBr and pressed into a pellet before testing. Finally, the crystallinity of ultra-fine CA fibers and sericin-AgNCs incorporated into ultra-fine CA fibers were analyzed using X-ray diffraction (XRD) model D/max-IIB, Rigaku. Samples were analyzed in the range of 10–50° at the scanning speed of $2\theta = 4°/min$. The ultra violet adsorption and fluorescence studies were evaluated using UV-visible spectrophotometer (Shimadzu UV2450, Kyoto, Japan) and UV-visible fluorescence spectrofluorometer (FLS920, Edinburgh, Livingston, UK) respectively.

3. Results and Discussion

3.1. Synthesis of Sericin-AgNCs

The synthesis of sericin-AgNCs was accomplished through the one pot, green synthesis route. As illustrated in Figure 1, sericin is actually a functional protein created by *Bombyx mori* (silkworms) in the production of silk, which make up the layers found on top of the fibrin. There are three different types of sericin, including sericin A, sericin B and sericin C. Due to the disulfide bonds and free cysteine, sericin play an important role in directing the synthesis of AgNCs. In the synthetic process, Ag ions were first introduced into the solution of sericin, where upon some Ag ions interact with the thiol groups of sericin to form functional Ag thiolate intermediates. After reduction, the spatial conformation of protein is further destructed from alkaline condition, followed by the exposure of more thiol site. The phenolic group of Y converts into a negative phenolic ion that can reduce Ag ions to Ag atoms. Finally, the resultant Ag atoms aggregate to form Ag clusters stabilized by the NH groups of the sericin [19].

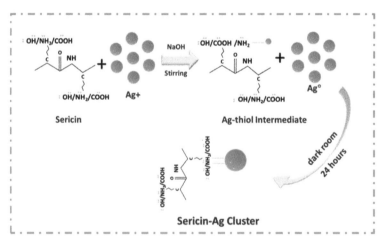

Figure 1. Typical illustration of synthesized sericin-AgNCs.

3.2. Physico-Chemical Analysis

The Figure 2A,B shows that sericin-AgNCs had maximum absorbance value at 440 nm and had fluorescence emission peaks at 500 nm respectively. Hence, the lyophilized sericin-AgNCs powder maintained the corresponding fluorescence emission. It is worth mentioned that samples were stable in their aqueous solutions and powder at 4 °C in the dark room for more than 3 months.

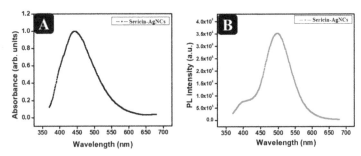

Figure 2. (**A**) UV-visible spectrum of sericin-AgNCs solution, and (**B**) UV-fluorescence spectrum of sericin-AgNCs solution.

The TEM technique was performed to assess the surface morphology and measure the size distribution of the sericin-AgNCs. Figure 3A shows the TEM image of sericin-AgNCs in which sericin-AgNCs got quasi-spherical and monodisperse having ultra-small size. Whereas the diameter distribution (Figure 3B) indicates that sericin-AgNCs have an average size of 2.5 ± 0.5 nm and obtained diameter distribution is in agreement with Gaussian fitting [9]. Moreover, the SEM images (Figure 3C) of ultrafine CA fibers revealed smooth surface fibrous morphology having average fiber diameter of 300 nm. After incorporation of sericin-AgNCs into ultrafine CA fibers, the surface morphology (Figure 3E) retains its smooth surface fibrous morphology with similar average fiber diameter of 300 nm due to very less amount of sericin-AgNCs incorporated into CA fibers [33].

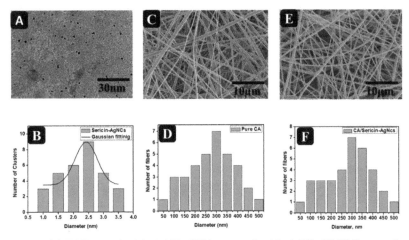

Figure 3. (**A**) TEM image of sericin-AgNCs, (**B**) histogram of sericin-AgNCs, (**C**) SEM images of pure CA fibers, (**D**) histogram of pure CA fibers, (**E**) SEM image of sericin-AgNCs incorporated into CA fibers, and (**F**) histogram of sericin-AgNCs incorporated into CA fibers.

The EDS was performed to confirm the presence of silver clusters in the ultrafine CA fibers after incorporating the sericin-AgNCs. Figure 4 shows the EDS elemental mapping and EDS spectrum of sericin-AgNCs incorporated ultrafine CA fibers. The EDS spectrum confirms the presence of silver clusters with 0.26% total weight of other components present in CA fibers. While the element mapping shows the uniform presence of silver clusters on the surface of CA fibers [34].

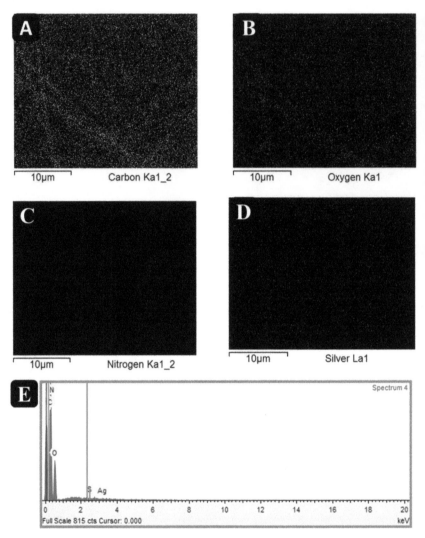

Figure 4. (**A–D**) EDS elemental mapping of sericin-AgNCs incorporated into ultrafine CA fibers, and (**E**) EDS spectrum and table (inset) showing the elemental composition of sericin-AgNCs incorporated into ultrafine CA fibers.

Figure 5 indicates XRD spectrum of electrospun CA fibers and incorporated sericin-AgNCs with CA fibers. The XRD structure of electrospun CA fibers membrane showed a wide peak at 23.2°, which explains an amorphous structure. While incorporating sericin-AgNCs, the diffractions peak shifted to about 20.6°, the comparative peak intensity showed a slightly decreases [35,36]. These results ensured the successful incorporation of sericin-AgNCs with in the electrospun CA fibers and explained that the AgNCs were actively interacting with the CA microstructure acting as nuclei sites during the spinning process to induce CA fibers crystallization.

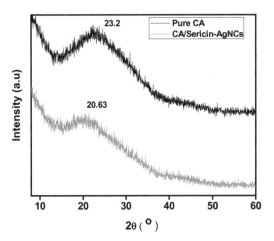

Figure 5. XRD results of pure CA and CA/sericin-AgNCs composite.

FTIR spectrum was used to analyze the chemical variations of electrospun CA fibers after incorporated sericin-AgNCs and confirms the successful hydrolysis of sericin-AgNCs with in the CA fibers. Figure 6 shows the broad spectrum of pure CA fibers and CA fibers with sericin-AgNCs. The pure CA fibers showed main typical peaks at 1745 cm^{-1} (C=O), 1375 cm^{-1} (C-CH$_3$) and 1227 cm^{-1} (C-O-C); corresponding to the vibrations of the acetate group of the CA fibers [37]. Furthermore, sericin-AgNCs incorporated into CA fibers showed a wide peak generated at 3345 cm^{-1} which corresponds the NH and OH groups of sericin [38], while the hydroxyl vibration at 3680 cm^{-1} was slightly decreased. In addition, a weak adsorption was found at 875 cm^{-1}, which corresponded to the silver compound and showed the successful incorporation of sericin-AgNCs into the CA fibers [36].

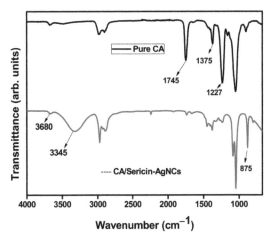

Figure 6. FTIR results of pure CA and sericin-AgNCs/CA fibers and both curves are shifted vertically for clarity.

3.3. Antibacterial Activity Assessment

The potential application of antibacterial activity on ultrafine CA fibers incorporated with sericin-AgNCs was examined against *E. coli* and *S. aureus*. As prepared electrospun CA fibers without sericin-AgNCs were examined as a control experiment. The bacterial colonies were incubated in a growing medium with the presence of various amount of sericin-AgNCs incorporated into ultrafine CA fibers and the antibacterial characteristics of resultant products were examined by standard shake flask protocol. Increasing the amount of sericin-AgNCs into CA fibers showed significant improvement in antibacterial activity. The total density of bacterial colonies for pure CA fibers was very high (Figure 7A,B), and antibacterial activity was observed nearly zero, while the small amount of sericin-AgNCs incorporated into CA fibers showed a remarkable improvement against *E. coli* and *S. aureus*. Although the amount of sericin-AgNCs was only 0.17 mg/mL ultrafine CA fibers, the antibacterial rate against *E. coli* and *S. aureus* could reach more than 85% antibacterial reduction (Figure 8). Furthermore, the antibacterial effect of sericin-AgNCs mediated ultrafine CA fibers was significantly improved with increase amount of sericin-AgNCs. The results indicated that the 1.7 mg/mL amount of sericin-AgNCs was enough to get more than 99.9% antibacterial reduction rate (Figure 7A,B). Moreover, the further increased amount of sericin-AgNCs showed similar antibacterial reduction rate with 1.7 mg/mL sericin-AgNCs and results were calculated in Figure 8. These results suggest that pure CA fiber did not show any antibacterial reduction and it was believed that due to addition of sericin-AgNCs into ultrafine CA fibers was the reason behind of this remarkable antibacterial reduction. Additionally, literature confirms that the antibacterial reduction of any product containing Ag was mainly verified by the total release rate of Ag ions [19]. In present work, due to ultrasmall particle size, the increase of Ag amount will lead a greater number of silver ions to enlargement and increase the total surface area of Ag nanoparticle; hence the release rate of silver ions was enhanced which can ultimately enhance the antibacterial reduction rate using sericin-AgNCs/CA composite. As per our experimental outcomes, it can be summarized that the CA fibers containing 1.7 mg/mL sericin-AgNCs showed excellent antibacterial reduction rate and amount of Ag was much lower than used in previous Ag/composites reports [5,8,36,38,39]. While promising characteristics of sericin makes our product very potential against numerous applications including biomedical materials, sports apparatus, and laminating films.

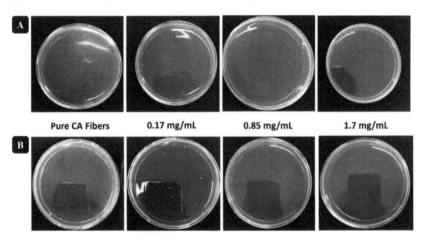

Figure 7. Bacterial colonies in the plates for (**A**) *Escherichia coli* and (**B**) *Staphylococcus aureus* at different contents of the sericin-AgNCs loaded into ultrafine CA fibers in the nutrient broth.

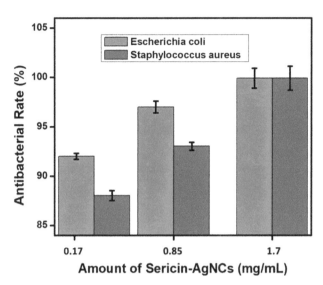

Figure 8. Degree of growth inhibition of different contents of sericin-AgNCs for bacterial suspension against *Escherichia coli* and *Staphylococcus aureus*.

It is worth noting that silver nanoclusters compared to other metals were considered safer antimicrobial agents due to their low toxicity. Sericin-AgNCs showed remarkable antibacterial property with ultra low amount of 1.7 mg/mL and results suggested that sericin-AgNCs incorporated into ultrafine CA fibers can be potential candidate for many biomedical applications including wound healing. To address potential impact of sericin-AgNCs, more and more research should be carried out to highlight its side effects on human health and ecology. This will lead to a better understanding and facilitate use of silver nanoclusters on commercial large scale as antimicrobial agents for applications onto different biomedical applications.

4. Conclusions

In the present study, a simple green synthesis route for sericin-AgNCs has been introduced and successfully incorporated into ultrafine CA fibers via electrospinning technique in order to report antibacterial properties. The average particle size of sericin-AgNCs was measured 2.5 nm and EDS results exhibited successful incorporation of sericin-AgNCs into ultrafine CA fibers. FTIR results confirmed the presence of sericin-AgNCs incorporated within ultrafine CA fibers. The antibacterial activities of sericin-AgNCs mediated ultrafine CA fibers were examined by shake flask method against the *E. coli* and *S. aureus* protocols. The results showed that sericin-AgNCs loaded ultrafine CA fibers exhibits outstanding antibacterial properties even with a ultralow content of sericin-AgNCs; as inhibition rate attained more than 90% at 0.17 mg/mL loading of sericin-AgNCs into ultrafine CA fibers and was finally reached >99.9% at 1.7 mg/mL amount of sericin-AgNCs against *E. coli* and *S. aureus*. The current results confirmed that the synergistic antibacterial effects of sericin-AgNCs with ultrafine CA fibers, and the composite can be used as promising antibacterial material for further study in biomedical applications.

Author Contributions: Conceptualization, M.M., H.Q. and P.G.; methodology, M.M. and H.Q.; formal analysis, R.F.Q., M.Y. and B.D.; investigation, S.H. and B.D.; writing—original draft preparation, S.H. and P.G.; writing—review and editing, M.M., R.F.Q. and Z.K.; supervision, Z.K. All authors have read and agreed to the published version of the manuscript.

Funding: This research received no external funding.

Institutional Review Board Statement: Not applicable.

Informed Consent Statement: Not applicable.

Data Availability Statement: The data presented in this study are available on request from the corresponding author.

Acknowledgments: This work was supported by Mehran University of Engineering and Technology Jamshoro, Pakistan and Soochow University, China.

Conflicts of Interest: The authors declare no conflict of interest.

References

1. Naeimi, A.; Payandeh, M.; Ghara, A.R.; Ghadi, F.E. In vivo evaluation of the wound healing properties of bio-nanofiber chitosan/polyvinyl alcohol incorporating honey and Nepeta dschuparensis. *Carbohydr. Polym.* **2020**, *240*, 116315. [CrossRef] [PubMed]
2. Shefa, A.A.; Sultana, T.; Park, M.K.; Lee, S.Y.; Gwon, J.-G.; Lee, B.-T. Curcumin incorporation into an oxidized cellulose nanofiber-polyvinyl alcohol hydrogel system promotes wound healing. *Mater. Des.* **2020**, *186*, 108313. [CrossRef]
3. Saratale, R.G.; Ghodake, G.S.; Shinde, S.K.; Cho, S.-K.; Saratale, G.D.; Pugazhendhi, A.; Bharagava, R.N. Photocatalytic activity of CuO/Cu(OH)$_2$ nanostructures in the degradation of Reactive Green 19A and textile effluent, phytotoxicity studies and their biogenic properties (antibacterial and anticancer). *J. Environ. Manag.* **2018**, *223*, 1086–1097. [CrossRef] [PubMed]
4. Saratale, R.G.; Shin, H.-S.; Kumar, G.; Benelli, G.; Ghodake, G.S.; Jiang, Y.Y.; Kim, D.S.; Saratale, G.D. Exploiting fruit byproducts for eco-friendly nanosynthesis: Citrus × clementina peel extract mediated fabrication of silver nanoparticles with high efficacy against microbial pathogens and rat glial tumor C6 cells. *Environ. Sci. Pollut. Res.* **2017**, *25*, 10250–10263. [CrossRef]
5. Matsumura, Y.; Yoshikata, K.; Kunisaki, S.-I.; Tsuchido, T. Mode of Bactericidal Action of Silver Zeolite and Its Comparison with That of Silver Nitrate. *Appl. Environ. Microbiol.* **2003**, *69*, 4278–4281. [CrossRef]
6. Saratale, R.G.; Saratale, G.D.; Ghodake, G.; Cho, S.-K.; Kadam, A.; Kumar, G.; Jeon, B.-H.; Pant, D.; Bhatnagar, A.; Shin, H.S. Wheat straw extracted lignin in silver nanoparticles synthesis: Expanding its prophecy towards antineoplastic potency and hydrogen peroxide sensing ability. *Int. J. Biol. Macromol.* **2019**, *128*, 391–400. [CrossRef]
7. Abduraimova, A.; Molkenova, A.; Duisembekova, A.; Mulikova, T.; Kanayeva, D.; Atabaev, T. Cetyltrimethylammonium Bromide (CTAB)-Loaded SiO$_2$–Ag Mesoporous Nanocomposite as an Efficient Antibacterial Agent. *Nanomaterials* **2021**, *11*, 477. [CrossRef]
8. Sheikh, F.A.; Barakat, N.A.M.; Kanjwal, M.A.; Chaudhari, A.A.; Jung, I.-H.; Lee, J.H.; Kim, H.Y. Electrospun antimicrobial polyurethane nanofibers containing silver nanoparticles for biotechnological applications. *Macromol. Res.* **2009**, *17*, 688–696. [CrossRef]
9. Saratale, G.D.; Saratale, R.G.; Kim, D.-S.; Kim, D.-Y.; Shin, H.S. Exploiting Fruit Waste Grape Pomace for Silver Nanoparticles Synthesis, Assessing Their Antioxidant, Antidiabetic Potential and Antibacterial Activity Against Human Pathogens: A Novel Approach. *Nanomaterials* **2020**, *10*, 1457. [CrossRef]
10. Chae, H.H.; Kim, B.-H.; Yang, K.S.; Rhee, J.I. Synthesis and antibacterial performance of size-tunable silver nanoparticles with electrospun nanofiber composites. *Synth. Met.* **2011**, *161*, 2124–2128. [CrossRef]
11. Panáček, A.; Kvítek, L.; Prucek, R.; Kolář, M.; Večeřová, R.; Pizúrová, N.; Sharma, V.K.; Nevěčná, T.; Zbořil, R. Silver Colloid Nanoparticles: Synthesis, Characterization, and Their Antibacterial Activity. *J. Phys. Chem. B* **2006**, *110*, 16248–16253. [CrossRef]
12. Pal, S.; Tak, Y.K.; Song, J.M. Does the Antibacterial Activity of Silver Nanoparticles Depend on the Shape of the Nanoparticle? A Study of the Gram-Negative Bacterium Escherichia coli. *Appl. Environ. Microbiol.* **2007**, *73*, 1712–1720. [CrossRef]
13. Yang, X.; Gondikas, A.P.; Marinakos, S.M.; Auffan, M.; Liu, J.; Hsu-Kim, H.; Meyer, J.N. Mechanism of Silver Nanoparticle Toxicity Is Dependent on Dissolved Silver and Surface Coating inCaenorhabditis elegans. *Environ. Sci. Technol.* **2011**, *46*, 1119–1127. [CrossRef]
14. El Badawy, A.M.; Silva, R.G.; Morris, B.; Scheckel, K.G.; Suidan, M.T.; Tolaymat, T.M. Surface Charge-Dependent Toxicity of Silver Nanoparticles. *Environ. Sci. Technol.* **2011**, *45*, 283–287. [CrossRef]
15. Gao, P.; Chang, X.; Zhang, D.; Cai, Y.; Chen, G.; Wang, H.; Wang, T.; Kong, T. Synergistic integration of metal nanoclusters and biomolecules as hybrid systems for therapeutic applications. *Acta Pharm. Sin. B* **2020**. [CrossRef]
16. Wang, B.; Zhao, M.; Mehdi, M.; Wang, G.; Gao, P.; Zhang, K.-Q. Biomolecule-assisted synthesis and functionality of metal nanoclusters for biological sensing: A review. *Mater. Chem. Front.* **2019**, *3*, 1722–1735. [CrossRef]
17. Luo, Z.; Zheng, K.; Xie, J. Engineering ultrasmall water-soluble gold and silver nanoclusters for biomedical applications. *Chem. Commun.* **2014**, *50*, 5143–5155. [CrossRef]
18. Choi, H.; Ko, S.-J.; Choi, Y.; Joo, P.; Kim, T.; Lee, B.R.; Jung, J.-W.; Choi, H.J.; Cha, M.; Jeong, J.-R.; et al. Versatile surface plasmon resonance of carbon-dot-supported silver nanoparticles in polymer optoelectronic devices. *Nat. Photon* **2013**, *7*, 732–738. [CrossRef]
19. Jin, J.-C.; Wu, X.-J.; Xu, J.; Wang, B.-B.; Jiang, F.-L.; Liu, Y. Ultrasmall silver nanoclusters: Highly efficient antibacterial activity and their mechanisms. *Biomater. Sci.* **2016**, *5*, 247–257. [CrossRef]

20. Saratale, R.G.; Cho, S.-K.; Saratale, G.D.; Kadam, A.A.; Ghodake, G.S.; Kumar, M.; Bharagava, R.N.; Kumar, G.; Kim, D.S.; Mulla, S.I.; et al. A comprehensive overview and recent advances on polyhydroxyalkanoates (PHA) production using various organic waste streams. *Bioresour. Technol.* **2021**, *325*, 124685. [CrossRef]
21. Chao, S.; Li, Y.; Zhao, R.; Zhang, L.; Li, Y.; Wang, C.; Li, X. Synthesis and characterization of tigecycline-loaded sericin/poly (vinyl alcohol) composite fibers via electrospinning as antibacterial wound dressings. *J. Drug Deliv. Sci. Technol.* **2018**, *44*, 440–447. [CrossRef]
22. Zhao, R.; Li, X.; Sun, B.; Zhang, Y.; Zhang, D.; Tang, Z.; Chen, X.; Wang, C. Electrospun chitosan/sericin composite nanofibers with antibacterial property as potential wound dressings. *Int. J. Biol. Macromol.* **2014**, *68*, 92–97. [CrossRef]
23. Khan, M.Q.; Kharaghani, D.; Shahzad, A.; Duy, N.P.; Hasegawa, Y.; Lee, J.; Kim, I.S. Fabrication of Antibacterial Nanofibers Composites by Functionalizing the Surface of Cellulose Acetate Nanofibers. *ChemistrySelect* **2020**, *5*, 1315–1321. [CrossRef]
24. Mahar, F.K.; He, L.; Wei, K.; Mehdi, M.; Zhu, M.; Gu, J.; Zhang, K.; Khatri, Z.; Kim, I. Rapid adsorption of lead ions using porous carbon nanofibers. *Chemosphere* **2019**, *225*, 360–367. [CrossRef]
25. Mehdi, M.; Mahar, F.K.; Qureshi, U.A.; Khatri, M.; Khatri, Z.; Ahmed, F.; Kim, I.S. Preparation of colored recycled polyethylene terephthalate nanofibers from waste bottles: Physicochemical studies. *Adv. Polym. Technol.* **2018**, *37*, 2820–2827. [CrossRef]
26. Son, W.K.; Youk, J.H.; Park, W.H. Antimicrobial cellulose acetate nanofibers containing silver nanoparticles. *Carbohydr. Polym.* **2006**, *65*, 430–434. [CrossRef]
27. Khatri, M.; Ahmed, F.; Ali, S.; Mehdi, M.; Ullah, S.; Duy-Nam, P.; Khatri, Z.; Kim, I.S. Photosensitive nanofibers for data recording and erasing. *J. Text. Inst.* **2021**, *112*, 429–436. [CrossRef]
28. Saquing, C.D.; Manasco, J.L.; Khan, S.A. Electrospun Nanoparticle-Nanofiber Composites via a One-Step Synthesis. *Small* **2009**, *5*, 944–951. [CrossRef]
29. Yang, Y.; Zhang, Z.; Wan, M.; Wang, Z.; Zou, X.; Zhao, Y.; Sun, L. A Facile Method for the Fabrication of Silver Nanoparticles Surface Decorated Polyvinyl Alcohol Electrospun Nanofibers and Controllable Antibacterial Activities. *Polymers* **2020**, *12*, 2486. [CrossRef]
30. Khalil, K.A.; Fouad, H.; Elsarnagawy, T.; Almajhdi, F.N. Preparation and characterization of electrospun PLGA/silver composite nanofibers for biomedical applications. *Int. J. Electrochem. Sci.* **2013**, *8*, 3483–3493.
31. Khatri, M.; Ahmed, F.; Shaikh, I.; Phan, D.-N.; Khan, Q.; Khatri, Z.; Lee, H.; Kim, I.S. Dyeing and characterization of regenerated cellulose nanofibers with vat dyes. *Carbohydr. Polym.* **2017**, *174*, 443–449. [CrossRef] [PubMed]
32. Kim, C.H.; Kim, S.Y.; Choi, K.S. Synthesis and Antibacterial Activity of Water-soluble Chitin Derivatives. *Polym. Adv. Technol.* **1997**, *8*, 319–325. [CrossRef]
33. Liang, S.; Zhang, G.; Min, J.; Ding, J.; Jiang, X. Synthesis and Antibacterial Testing of Silver/Poly (Ether Amide) Composite Nanofibers with Ultralow Silver Content. *J. Nanomater.* **2014**, *2014*, 1–10. [CrossRef]
34. Hussain, N.; Mehdi, M.; Yousif, M.; Ali, A.; Ullah, S.; Siyal, S.H.; Hussain, T.; Kim, I. Synthesis of Highly Conductive Electrospun Recycled Polyethylene Terephthalate Nanofibers Using the Electroless Deposition Method. *Nanomaterials* **2021**, *11*, 531. [CrossRef]
35. Jiang, G.-H.; Wang, L.; Chen, T.; Yu, H.-J.; Wang, J.-J. Preparation and characterization of dendritic silver nanoparticles. *J. Mater. Sci.* **2005**, *40*, 1681–1683. [CrossRef]
36. Xu, F.; Weng, B.; Materon, L.A.; Kuang, A.; Trujillo, J.A.; Lozano, K. Fabrication of cellulose fine fiber based membranes embedded with silver nanoparticles via Forcespinning. *J. Polym. Eng.* **2016**, *36*, 269–278. [CrossRef]
37. Ahmed, F.; Arbab, A.A.; Jatoi, A.W.; Khatri, M.; Memon, N.; Khatri, Z.; Kim, I.S. Ultrasonic-assisted deacetylation of cellulose acetate nanofibers: A rapid method to produce cellulose nanofibers. *Ultrason. Sonochem.* **2017**, *36*, 319–325. [CrossRef]
38. Hadipour-Goudarzi, E.; Montazer, M.; Latifi, M.; Aghaji, A.A.G. Electrospinning of chitosan/sericin/PVA nanofibers incorporated with in situ synthesis of nano silver. *Carbohydr. Polym.* **2014**, *113*, 231–239. [CrossRef]
39. Islam, M.S.; Yeum, J.H.J.C.; Physicochemical, S.A.; Aspects, E. Electrospun pullulan/poly (vinyl alcohol)/silver hybrid nanofibers: Preparation and property characterization for antibacterial activity. *Colloids Surf. A Physicochem. Eng. Asp.* **2013**, *436*, 279–286. [CrossRef]

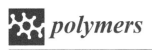

MDPI

Article

Fabrication of Poly(Ethylene-glycol 1,4-Cyclohexane Dimethylene-Isosorbide-Terephthalate) Electrospun Nanofiber Mats for Potential Infiltration of Fibroblast Cells

Sofia El-Ghazali [1,2], Muzamil Khatri [2], Mujahid Mehdi [3], Davood Kharaghani [4], Yasushi Tamada [5], Anna Katagiri [5], Shunichi Kobayashi [1,*] and Ick Soo Kim [2,*]

[1] Department of Biomedical Engineering, Division of Biomedical Engineering, Faculty of Science and Technology, Shinshu University, Tokida 3-15-1, Ueda, Nagano Prefecture 386-8567, Japan; elghazalisofia@gmail.com
[2] Nano Fusion Technology Research Group, Division of Frontier Fibers, Institute for Fiber Engineering (IFES), Interdisciplinary Cluster for Cutting Edge Research (ICCER), Shinshu University, Tokida 3-15-1, Ueda, Nagano Prefecture 386-8567, Japan; muzamilkhatri@gmail.com
[3] National Engineering Laboratory for Modern Silk, College of Textile and Clothing Engineering, Soochow University, Suzhou 215123, China; mujahid11te83@gmail.com
[4] Department of Calcified Tissue Biology, Graduate School of Biomedical and Health Sciences, Hiroshima University, 1-2-3 Kasumi, Minami-Ku, Hiroshima 734-8553, Japan; Kharaghani66@gmail.com
[5] Faculty of Textile Science and Technology Bioresource and Environmental Science, Shinshu University, Tokida 3-15-1, Ueda, Nagano Prefecture 386-8567, Japan; ytamada@shinshu-u.ac.jp (Y.T.); 19fs703j@shinshu-u.ac.jp (A.K.)
* Correspondence: shukoba@shinshu-u.ac.jp (S.K.); kim@shinshu-u.ac.jp (I.S.K.); Tel.: +81-268-21-5511 (S.K.); +81-268-21-5439 (I.S.K.)

Citation: El-Ghazali, S.; Khatri, M.; Mehdi, M.; Kharaghani, D.; Tamada, Y.; Katagiri, A.; Kobayashi, S.; Kim, I.S. Fabrication of Poly(Ethylene-glycol 1,4-Cyclohexane Dimethylene-Isosorbide-Terephthalate) Electrospun Nanofiber Mats for Potential Infiltration of Fibroblast Cells. Polymers 2021, 13, 1245. https://doi.org/10.3390/polym13081245

Academic Editor: Marián Lehocký

Received: 27 March 2021
Accepted: 9 April 2021
Published: 12 April 2021

Publisher's Note: MDPI stays neutral with regard to jurisdictional claims in published maps and institutional affiliations.

Abstract: Recently, bio-based electrospun nanofiber mats (ENMs) have gained substantial attention for preparing polymer-based biomaterials intended for use in cell culture. Herein, we prepared poly(ethylene-glycol 1,4-Cyclohexane dimethylene-isosorbide-terephthalate) (PEICT) ENMs using the electrospinning technique. Cell adhesion and cell viability of PEICT ENMs were checked by fibroblast cell culture. Field emission electron microscope (FE-SEM) image showed a randomly interconnected fiber network, smooth morphology, and cell adhesion on PEICT ENM. Fibroblasts were cultured in an adopted cell culturing environment on the surface of PEICT ENMs to confirm their biocompatibility and cell viability. Additionally, the chemical structure of PEICT ENM was checked under Fourier-transform infrared (FTIR) spectroscopy and the results were supported by -ray photoelectron (XPS) spectroscopy. The water contact angle (WCA) test showed the hydrophobic behavior of PEICT ENMs in parallel to good fibroblast cell adhesion. Hence, the results confirmed that PEICT ENMs can be potentially utilized as a biomaterial.

Keywords: nanofibers; fibroblast; cell culture; cell adhesion; cell viability; biobased polyester

1. Introduction

Recently, bio-based polyesters have been widely studied for biomedical applications such as cell culture, drug delivery, and wound dressing [1–6]. Scientists have started exploring the biomedical applications of isosorbide-based polyesters by virtue of their high tensile strength and good thermal stability [4,6–9]. Electrospun nanofiber mats (ENMs), due to their high surface area, ease of production, fiber network interconnection, biocompatibility, breathability, and resemblance to natural extracellular matrix have been potentially considered for cell culture applications viz cell adhesion, cell migration, and cell viability [10–13]. In parallel, ENMs have been utilized in several applications including sensors [14], wound dressing [10], drug delivery [5], tissue engineering [15], smart apparels [16,17] high performance filters (water, bacteria, virus, heavy metal, and such impurities) [18–21]. Depending on the required application, cell adhesion and cell migration behaviors may differ due to the difference in cell size, type, or the culturing surface [4,6]. Scaffolds based on ENMs possess good biodegradability, biocompatibility, non-thrombogenicity, and non-immunogenicity, however, the optimization of such materials is challenging in terms of cell viability, cell infiltration, cell adhesion, and migration properties [4–6,10,11,22].

Poly(ethylene-glycol 1,4-Cyclohexane dimethylene-isosorbide-terephthalate) (PEICT) ENMs, in light of their high glass transition temperature [23], good mechanical properties [8], flexibility [6], thermal stability [4], and good crystallinity [23] have potentially been studied for various biomedical applications [4,6].

In our previous report, we investigated pristine poly(1,4-Cyclohexane di-methylene-isosorbide-terephthalate) (PICT), PEICT, and their blended composition. We observed that no sufficient cell infiltration could be achieved due to the unsuitable nanofiber network on the surface, which could not hold breast epithelial cells on the surface of PEICT ENMs [4], therefore, we tried to explore the potential of PEICT ENMs for cell culture applications and to check the effect of surface properties of PEICT ENMs whether they allow fibroblast cell infiltration [4]. Fibroblast cells have been used to check the biocompatibility, cell infiltration, cell adhesion, and cell viability of different polymeric materials [6,13].

To serve the main purpose of this research, isosorbide biobased PEICT ENMs were prepared via electrospinning, and used as a fibroblast cell permeable layer for the first time. There are a few reports that have discussed the chemical structure of PEICT only characterized by FTIR spectroscopy, where in this report, the chemical structure of PEICT ENMs was significantly explored by x-ray photoelectron spectroscopy (XPS) validated by the FTIR results [4,8,23].

The wettability of PEICT ENMs was checked by the WCA test with respect to time as one of the important factors for cell adhesion to optimize the initial resting of fibroblast cells on PEICT ENMs. The nanofiber network and surface topography of PEICT ENMs were studied with respect to fibroblast cell infiltration. The results revealed the potential of PEICT ENM to be used as a biomaterial for fibroblast cell culture.

2. Materials and Methods

2.1. Materials

The polymer PEICT with an average molecular weight of 46,800 (g/mol) was supplied by SK Chemicals, Gyeonggi-do, Korea [8]. Chloroform (99%), and trifluoroacetic acid (TFA) (98%) were purchased from Wako, Pure Chemical Industries Ltd., Osaka, Japan. Ethanol (99.8%) was obtained by Merck, Japan. Cell migration on PEICT ENMs was performed using NIH 3T3 mouse fibroblast obtained from ATCC (Manassas, VA), USA [24]. The 24-well cell culture flasks and Petri-dishes were obtained from SPL Life Sciences (Pocheon, Korea). Ethylenediaminetetraacetic acid (EDTA) was purchased from Merck (KGaA), Germany. Dulbecco's modified Eagle's medium (DMEM) was supplied by Nacalai Tesque, Kyoto, Japan. Fetal bovine serum (FBS) and Dulbecco's phosphate-buffered saline (DPBS) were provided by HyClone (Logan, UT, USA). The Cell Counting Kit-8 (CCK-8) reagent was obtained from Dojindo Laboratories Japan [25].

2.2. Preparation of PEICT ENM

A total of 11% (w/w) of PEICT solution was stirred at room temperature in chloroform:TFA with a ratio of 3:1 until the formation of a transparent polymer solution. A 10 mL syringe filled with the PEICT polymer solution was installed on an electrospinning power supply (Har-100*12, Matsusada Co., Tokyo, Japan) and supplied with a high voltage of 12.5 kV, keeping a distance of 18 cm from the electrospinning tip to the collector. PEICT ENMs were dried in air at room temperature prior to other necessary characterizations.

2.3. Characterizations

The in vitro cytocompatibility of PEICT ENMs was analyzed by NIH 3T3 mouse fibroblast cells, DMEM was used with 10% fetal bovine serum (FBS) incubated at 37 °C with 5% CO_2. The PEICT ENMs were punched into 3 mm disc-shaped webs and poured into 70% ethanol for sterilization, then washed using PBS after 30 min of dipping. The sterilized PEICT ENMs were placed in a culture flask. Subsequently, each cell culturing well was seeded with NIH 3T3 cells at (1×10^3) density. The cell culture assessment on PEICT nanofibers was carried for one, three, and seven days, where 10 μL WST-1 was added to each flask intermittently after two days. The toxicity, cell proliferation, cell viability, and cell adhesion properties of PEICT ENMs were studied in accordance with previously reported articles [22,24,25].

The morphology of PEICT ENMs was assessed under (field emission electron microscope (FE-SEM; JEOL JSM 6700 F) by JEOL, Japan with the accelerating voltage of (15 kV) and a maximum magnification of 2000×, whereas the morphology of PEICT ENMs before and after cell culture was assessed under SEM (JSM-6010LA) by JEOL, Japan with the accelerating voltage of (10 kV). A fin coater (JFC-1200) by JEOL, Japan was used for coating PEICT ENMs using Pb-Pt for 120 s. The average diameter histogram of PEICT ENMs was calculated using ImageJ software. To investigate the fibroblast cell adhesion and migration on PEICT ENMs, the morphology of fibroblast cells was assessed under scanning electron microscope (SEM) and the images were taken before and after seven days of cell culture. Fibroblast cells, after seven days of culture, were fixed on the PEICT ENMs with 4% of paraformaldehyde for 4 h at 4 °C. PEICT ENMs containing the adhered fibroblasts were dehydrated using ethanol and dried in a vacuum prior to taking images [25].

The chemical composition of PEICT ENMs was assessed under Attenuated Total Reflection Fourier-transform infrared spectroscopy (ATR-FTIR) mode using IR Prestige-21, Shimadzu (Kyoto, Japan). Transmittance spectra were assessed in the wavelength range of 670 to 3000 cm^{-1} at room temperature.

The presence of elements available in the PEICT ENMs was checked under XPS supplied from AXIS Ultra (Schimadzu) with a hemispherical sector analyzer (HSA), dual-anode x-ray source Al/Mg, and the detector. A MgKα x-ray source with 1253.6 eV was used at 1.4×10^{-9} torr pressure.

3. Results and Discussion

3.1. Fibroblast Cell Culture, Cell Adhesion, and Cell Viability on PEICT ENMs

Fibroblast cells were cultured on PEICT ENMs to check the biocompatibility of these nanofibrous mats [22]. Figure 1A shows the microscopic images of the fibroblast cells in the culture flask for one day, three days, and seven days [25]. SEM images were taken after seven days of cell culture on PEICT ENMs, which revealed good cell adhesion as shown in Figure 1B. The reason for this may be the smooth morphology of PEICT ENMs and the optimum cellular infiltration into the porous nanofiber mats. The bar graph shown in Figure 1C reveals cell viability, which demonstrates that the presence of PEICT ENMs in the culturing environment does not have much negative influence on cell growth, where 85–90% of the healthy cells can survive up to one day (until the first day) of culture. After three days of culture, the number of cells was decreased by 35%, but after seven days, the remaining cells showed very good proliferation and the fibroblast cells had multiplied and reached 70% in parallel to very good cell adhesion. These findings suggest that PEICT ENMs can potentially be considered for fibroblast cell culture application.

3.2. Morphology of PEICT ENMs

To assess the optimized network of nanofibers for the culture and the infiltration of fibroblast cells on the surface of PEICT ENMs, it was necessary to check the morphology and the average fiber diameter of PEICT ENMs. Figure 2A shows the bead-free and the smooth morphology of PEICT nanofibers when electrospun with 11% polymer concentration on a rotating cylinder at a speed of 70 rpm. Figure 2B shows the average diameter distribution graph of PEICT ENMs. The average fiber diameter was found to be 550 ± 60 nm, and the average fiber dimeter distributions of PEICT nanofibers ranging from 300 nm to 700 nm, indicating a non-uniform filtration surface of the PEICT ENM layer, which may be a reason that the infiltration was not 100%.

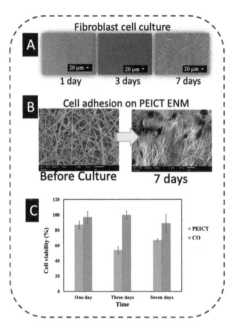

Figure 1. (**A**) Microscopic images of fibroblast cells in the culture flask. (**B**) SEM image of cell adhesion of PEICT ENM. (**C**) Cell viability of PEICT ENMs.

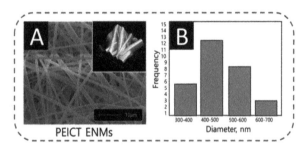

Figure 2. (**A**) FE-SEM image inset with AFM image, and (**B**) average diameter histogram of PEICT ENMs.

Compared to the previous reports on PEICT ENMs, this study showed a slightly smaller average fiber diameter size and less diameter distribution as exhibited in the average fiber diameter histogram shown in Figure 2B. In addition, the fiber topography on the surface of PEICT ENM, revealed by the inset atomic force microscopic (AFM) image of Figure 2A, showed a smaller number of gaps between the nanofibers. Good fibroblast cell adhesion on the PEICT ENM surface was perhaps due to the similar frequency of the nanofiber's average diameter, which helped in providing smoothness in parallel to an interconnected nanofiber network to support fibroblast cell infiltration [4,26–29].

3.3. FTIR Spectroscopy of PEICT ENMs

Figure 3 shows the FTIR spectrum of PEICT ENM; the intensive cis peak appearing at 1244 cm^{-1} reveals the ethylene glycol present in its chemical structure, which is in agreement with previous reports [6,8,23]. FTIR spectroscopy further confirms the presence of CH at 2930 cm^{-1} and 726 cm^{-1}, C=O at 1715 cm^{-1} and CN at 1094 cm^{-1}. The abundance of the –CH group on the surface of PEICT ENM may also be the reason for fibroblast cell growth and cell adhesion [29,30].

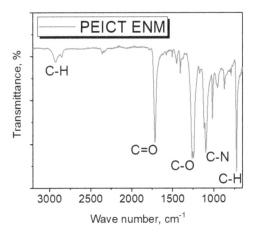

Figure 3. FTIR spectrum of PEICT ENM.

3.4. XPS Spectroscopy of PEICT ENMs

The main elements present in PEICT ENMs are isosorbide and ethylene glycol, which were previously reported in published articles and revealed by FTIR spectroscopy [4,6,31]. For a significant chemical composition assessment, we chose XPS spectroscopy to deeply understand the chemical composition of PEICT ENMs. Figure 4 shows the XPS spectra of PEICT ENMs. Figure 4A shows the wide XPS spectrum of a PEICT ENM, where two main peaks can be found in the range of Carbon 1s (C1s) and Oxygen 1s (O1s). Therefore, we took XPS spectroscopy of these two intensive regions, C1s ranging from 282 eV to 289.5 eV and O1s ranging from 528 eV and 534 eV, while no intensive peak was observed at N1s due to the small amount of the CN group present in the chemical structure of PEICT ENMs. Figure 4B demonstrates the XPS spectra in C1s, confirming the –CH group abundance on the surface of the PEICT ENM, and also shows the presence of O=C–O, CN, C–C, C=O, and C–O, demonstrating the chemical composition of PEICT ENM more significantly than FTIR and other previous reports [23,31–34]. The XPS spectra of O1s is given in Figure 4C, indicating C=O and C–O present in the chemical composition of PEICT ENM, which validates the results of C1s and FTIR [34,35].

3.5. Wetting Properties of PEICT ENMs

In this report, we cultured fibroblast cells on PEICT nanofibers for the very first time, therefore, it was necessary to check the surface wetting properties as one of the important parameters for cell adhesion [27,30]. PEICT ENMs were assessed by the water contact angle (WCA) test [18,36]. Compared to the previous reports on the WCA of PEICT ENMs (114°) [4], the contact angle observed in this study was slightly higher than 120°, where the obvious reason for this was the similar frequency of the nanofibers' average diameter, which provided a regular surface with finer fibers that led the water droplet being stable for a certain time prior to starting adsorption on PEICT ENMs. According to Figure 5, the water droplet took about 270 s to reach the angle 0° from 120°, which is in the range of hydrophobic surfaces. Even though the surface of PEICT ENM is hydrophobic [37], the cell infiltration observed was very good because of the interconnected nanofiber network with a smooth morphology, which also played an important role in cell adhesion. Additionally, the WCA test suggested that five minutes were enough as a resting period for fibroblast initial cell adherence on PEICT ENMs.

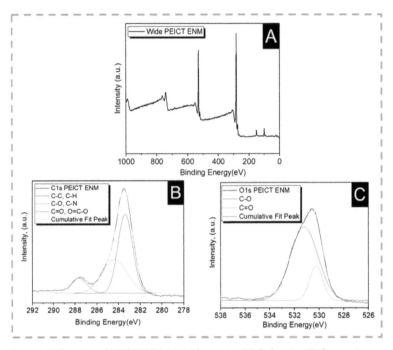

Figure 4. XPS spectra of PEICT ENM. (**A**) Wide spectra. (**B**) Carbon 1s. (**C**) Oxygen 1s.

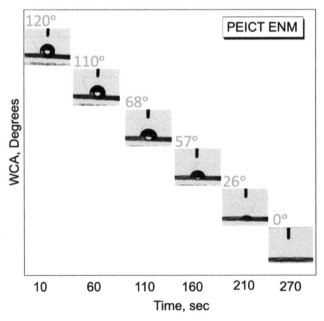

Figure 5. Water contact angle test of PEICT ENMs.

4. Conclusions

PEICT ENMs were successfully electrospun at 11% w/w polymer concentration, the electrospun PEICT nanofibers showed bead-free and smooth morphology with an average diameter of 550 ± 60 nm, having less diameter distributions compared to previous reports on electropun PEICT nanofibers. SEM images showed a good cell viability (70%) and cell adhesion properties on the surface of PEICT ENMs after seven days of cell culture. FTIR and XPS spectroscopies confirmed the chemical structure of PEICT ENMs and the bulk presence of the C–H group on its surface, which was one of the reasons for the good cell adhesion. The results manifested a dependency of the cell infiltration on the fiber network and the average diameter of PEICT ENMs. Ultimately, less gaps on the surface of PEICT ENMs may be a reason for cell infiltration. The WCA test revealed the hydrophobic behavior of the mats with an average WCA of 120 °, which demonstrated the optimum duration of 5 min for the initial resting of cells for their culture prior to starting their culture on PEICT ENMs. Therefore, the results confirmed the potential of PEICT ENMs for fibroblast cell culture application.

Author Contributions: Conceptualization, S.E.-G.; Methodology S.E.-G.; Software, S.E.-G.; Validation, I.S.K.; Formal analysis, Y.T., M.M. and A.K.; Investigation, D.K. and S.E.-G.; Resources, S.K.; Data curation, S.E.-G.; Writing—original draft preparation, S.E.-G. and M.K.; Writing—review and editing, S.E.-G. and S.K.; Visualization, S.E.-G.; Supervision, S.K. and I.S.K. All authors have read and agreed to the published version of the manuscript.

Funding: This work was supported by a Grant-in-Aid for the Shinshu University Advanced Leading Graduate Program by the Ministry of Education, Culture, Sports, Science, and Technology (MEXT), Japan.

Institutional Review Board Statement: Not applicable.

Informed Consent Statement: Not applicable.

Data Availability Statement: The data presented in this study are available on request from the corresponding author. There is no public repository at our institution.

Conflicts of Interest: The authors declare no conflict of interest.

References

1. Kasmi, N.; Pinel, C.; Perez, D.D.S.; Dieden, R.; Habibi, Y. Synthesis and characterization of fully biobased polyesters with tunable branched architectures. *Polym. Chem.* **2021**, *12*, 991–1001. [CrossRef]
2. Bansal, K.K.; Rosenholm, J.M. Synthetic polymers from renewable feedstocks: An alternative to fossil-based materials in biomedical applications. *Ther. Deliv.* **2020**, *11*, 297–300. [CrossRef] [PubMed]
3. Prabakaran, R.; Marie, J.M.; Xavier, A.J.M. Biobased Unsaturated Polyesters Containing Castor Oil-Derived Ricinoleic Acid and Itaconic Acid: Synthesis, In Vitro Antibacterial, and Cytocompatibility Studies. *ACS Appl. Bio Mater.* **2020**, *3*, 5708–5721. [CrossRef]
4. El-Ghazali, S.; Khatri, M.; Hussain, N.; Khatri, Z.; Yamamoto, T.; Kim, S.H.; Kobayashi, S.; Kim, I.S. Characterization and biocompatibility evaluation of artificial blood vessels prepared from pristine poly (Ethylene-glycol-co-1, 4-cyclohexane dimethylene-co-isosorbide terephthalate), poly (1, 4 cyclohexane di-methylene-co-isosorbide terephthalate) nanofibers and their blended composition. *Mater. Today Commun.* **2021**, *26*, 102113. [CrossRef]
5. Das, S.S.; Bharadwaj, P.; Bilal, M.; Barani, M.; Rahdar, A.; Taboada, P.; Bungao, S.; Kyzas, G.Z. Stimuli-Responsive Polymeric Nanocarriers for Drug Delivery, Imaging, and Theragnosis. *Polymers* **2020**, *12*, 1397. [CrossRef] [PubMed]
6. Khan, M.Q.; Lee, H.; Khatri, Z.; Kharaghani, D.; Khatri, M.; Ishikawa, T.; Im, S.S.; Kim, I.S. Fabrication and characterization of nanofibers of honey/poly (1, 4-cyclohexane dimethylene isosorbide trephthalate) by electrospinning. *Mater. Sci. Eng. C* **2017**, *81*, 247–251. [CrossRef]
7. Park, S.; Thanakkasaranee, S.; Shin, H.; Ahn, K.; Sadeghi, K.; Lee, Y.; Tak, G.; Seo, J. Preparation and characterization of heat-resistant PET/bio-based polyester blends for hot-filled bottles. *Polym. Test.* **2020**, *91*, 106823. [CrossRef]
8. Lee, H.; Koo, J.M.; Sohn, D.; Kim, I.S.; Im, S.S. High thermal stability and high tensile strength terpolyester nanofibers containing biobased monomer: Fabrication and characterization. *RSC Adv.* **2016**, *6*, 40383–40388. [CrossRef]
9. Legrand, S.; Jacquel, N.; Amedro, H.; Saint-Loup, R.; Pascault, J.P.; Rousseau, A.; Fenouillot, F. Synthesis and properties of poly (1, 4-cyclohexanedimethylene-co-isosorbide terephthalate), a biobased copolyester with high performances. *Eur. Polym. J.* **2019**, *115*, 22–29. [CrossRef]
10. Bui, H.T.; Chung, O.H.; Cruz, J.D.; Park, J.S. Fabrication and characterization of electrospun curcumin-loaded polycaprolactone-polyethylene glycol nanofibers for enhanced wound healing. *Macromol. Res.* **2014**, *22*, 1288–1296. [CrossRef]

11. Lanzalaco, S.; Del Valle, L.J.; Turon, P.; Weis, C.; Estrany, F.; Alemán, C.; Armelin, E. Polypropylene mesh for hernia repair with controllable cell adhesion/de-adhesion properties. *J. Mater. Chem. B* **2020**, *8*, 1049–1059. [CrossRef] [PubMed]
12. Cavalcanti-Adam, E.A. Building nanobridges for cell adhesion. *Nat. Mater.* **2019**, *18*, 1272–1273. [CrossRef] [PubMed]
13. Khatri, Z.; Jatoi, A.W.; Ahmed, F.; Kim, I.S. Cell adhesion behavior of poly (ε-caprolactone)/poly (L-lactic acid) nanofibers scaffold. *Mater. Lett.* **2016**, *171*, 178–181. [CrossRef]
14. Khatri, M.; Ahmed, F.; Ali, S.; Mehdi, M.; Ullah, S.; Duy-Nam, P.; Khatri, Z.; Kim, I.S. Photosensitive nanofibers for data recording and erasing. *J. Text. Inst.* **2021**, *112*, 429–436. [CrossRef]
15. Udomluck, N.; Lee, H.; Hong, S.; Lee, S.H.; Park, H. Surface functionalization of dual growth factor on hydroxyapatite-coated nanofibers for bone tissue engineering. *Appl. Surf. Sci.* **2020**, *520*, 146311. [CrossRef]
16. Khatri, M.; Ahmed, F.; Shaikh, I.; Phan, D.N.; Khan, Q.; Khatri, Z.; Lee, H.; Kim, I.S. Dyeing and characterization of regenerated cellulose nanofibers with vat dyes. *Carbohydr. Polym.* **2017**, *174*, 443–449. [CrossRef] [PubMed]
17. Khatri, M.; Ahmed, F.; Jatoi, A.W.; Mahar, R.B.; Khatri, Z.; Kim, I.S. Ultrasonic dyeing of cellulose nanofibers. *Ultrason. Sonochem.* **2016**, *31*, 350–354. [CrossRef]
18. Khatri, M.; Khatri, Z.; El-Ghazali, S.; Hussain, N.; Qureshi, U.A.; Kobayashi, S.; Ahmed, F.; Kim, I.S. Zein nanofibers via deep eutectic solvent electrospinning: Tunable morphology with super hydrophilic properties. *Sci. Rep.* **2020**, *10*, 1–11. [CrossRef] [PubMed]
19. Cheng, P.; Wang, X.; Liu, Y.; Kong, C.; Liu, N.; Wan, Y.; Guo, Q.; Liu, K.; Lu, Z.; Li, M.; et al. Ag nanoparticles decorated PVA-co-PE nanofiber-based membrane with antifouling surface for highly efficient inactivation and interception of bacteria. *Appl. Surf. Sci.* **2020**, *506*, 144664. [CrossRef]
20. Fahimirad, S.; Fahimirad, Z.; Sillanpää, M. Efficient removal of water bacteria and viruses using electrospun nanofibers. *Sci. Total Environ.* **2020**, *751*, 141673. [CrossRef]
21. Phan, D.N.; Khan, M.Q.; Nguyen, N.T.; Phan, T.T.; Ullah, A.; Khatri, M.; Kien, N.N.; Kim, I.S. A review on the fabrication of several carbohydrate polymers into nanofibrous structures using electrospinning for removal of metal ions and dyes. *Carbohydr. Polym.* **2020**, *252*, 117175. [CrossRef] [PubMed]
22. Ullah, A.; Saito, Y.; Ullah, S.; Haider, M.K.; Nawaz, H.; Duy-Nam, P.; Kharaghani, D.; Kim, I.S. Bioactive Sambong oil-loaded electrospun cellulose acetate nanofibers: Preparation, characterization, and in-vitro biocompatibility. *Int. J. Biol. Macromol.* **2021**, *166*, 1009–1021. [CrossRef] [PubMed]
23. Phan, D.N.; Lee, H.; Choi, D.; Kang, C.Y.; Im, S.S.; Kim, I.S. Fabrication of Two Polyester Nanofiber Types Containing the Biobased Monomer Isosorbide: Poly (Ethylene Glycol 1,4-Cyclohexane Dimethylene Isosorbide Terephthalate) and Poly (1,4-Cyclohexane Dimethylene Isosorbide Terephthalate). *Nanomaterials* **2018**, *8*, 56. [CrossRef]
24. Kharaghani, D.; Gitigard, P.; Ohtani, H.; Kim, K.O.; Ullah, S.; Saito, Y.; Kim, I.S. Design and characterization of dual drug delivery based on in-situ assembled PVA/PAN core-shell nanofibers for wound dressing application. *Sci. Rep.* **2019**, *9*, 1–11. [CrossRef]
25. Kharaghani, D.; Jo, Y.K.; Khan, M.Q.; Jeong, Y.; Cha, H.J.; Kim, I.S. Electrospun antibacterial polyacrylonitrile nanofiber membranes functionalized with silver nanoparticles by a facile wetting method. *Eur. Polym. J.* **2018**, *108*, 69–75. [CrossRef]
26. Mirtič, J.; Balažic, H.; Zupančič, Š.; Kristl, J. Effect of Solution Composition Variables on Electrospun Alginate Nanofibers: Response Surface Analysis. *Polymers* **2019**, *11*, 692. [CrossRef] [PubMed]
27. Han, D.G.; Ahn, C.B.; Lee, J.H.; Hwang, Y.; Kim, J.H.; Park, K.Y.; Lee, J.W.; Son, K.H. Optimization of Electrospun Poly(caprolactone) Fiber Diameter for Vascular Scaffolds to Maximize Smooth Muscle Cell Infiltration and Phenotype Modulation. *Polymers* **2019**, *11*, 643. [CrossRef] [PubMed]
28. Habib, S.; Zavahir, S.; Abusrafa, A.E.; Abdulkareem, A.; Sobolčiak, P.; Lehocky, M.; Vesela, D.; Humpolíček, P.; Popelka, A. Slippery Liquid-Infused Porous Polymeric Surfaces Based on Natural Oil with Antimicrobial Effect. *Polymers* **2021**, *13*, 206. [CrossRef]
29. Dong, X.; Yuan, X.; Wang, L.; Liu, J.; Midgley, A.C.; Wang, Z.; Wang, K.; Liu, J.; Zhu, M.; Kong, D. Construction of a bilayered vascular graft with smooth internal surface for improved hemocompatibility and endothelial cell monolayer formation. *Biomaterials* **2018**, *181*, 1–14. [CrossRef]
30. Arima, Y.; Iwata, H. Effect of wettability and surface functional groups on protein adsorption and cell adhesion using well-defined mixed self-assembled monolayers. *Biomaterials* **2017**, *28*, 3074–3082. [CrossRef]
31. Yoon, W.J.; Hwang, S.Y.; Koo, J.M.; Lee, Y.J.; Lee, S.U.; Im, S.S. Synthesis and Characteristics of a Biobased High-Tg Terpolyester of Isosorbide, Ethylene Glycol, and 1,4-Cyclohexane Dimethanol: Effect of Ethylene Glycol as a Chain Linker on Polymerization. *Macromolecules* **2013**, *46*, 7219–7231. [CrossRef]
32. Yu, Y.; Jiang, X.; Fang, Y.; Chen, J.; Kang, J.; Cao, Y.; Xiang, M. Investigation on the Effect of Hyperbranched Polyester Grafted Graphene Oxide on the Crystallization Behaviors of β-Nucleated Isotactic Polypropylene. *Polymers* **2019**, *11*, 1988. [CrossRef] [PubMed]
33. Narayanan, G.; Shen, J.; Boy, R.; Gupta, B.S.; Tonelli, A.E. Aliphatic Polyester Nanofibers Functionalized with Cyclodextrins and Cyclodextrin-Guest Inclusion Complexes. *Polymers* **2018**, *10*, 428. [CrossRef] [PubMed]
34. Zhang, W.; Zheng, C.; Zhang, Y.; Guo, W. Preparation and Characterization of Flame-Retarded Poly(butylene terephthalate)/Poly(ethylene terephthalate) Blends: Effect of Content and Type of Flame Retardant. *Polymers* **2019**, *11*, 1784. [CrossRef]

35. Mehdi, M.; Mahar, F.K.; Qureshi, U.A.; Khatri, M.; Khatri, Z.; Ahmed, F.; Kim, I.S. Preparation of colored recycled polyethylene terephthalate nanofibers from waste bottles: Physicochemical studies. *Adv. Polym. Technol.* **2018**, *37*, 2820–2827. [CrossRef]
36. Khatri, M.; Hussain, N.; El-Ghazali, S.; Yamamoto, T.; Kobayashi, S.; Khatri, Z.; Kim, I.S. Ultrasonic-assisted dyeing of silk fibroin nanofibers: An energy-efficient coloration at room temperature. *Appl. Nanosci.* **2020**, *10*, 917–930. [CrossRef]
37. Chen, C.Y.; Huang, S.Y.; Wan, H.Y.; Chen, Y.T.; Yu, S.K.; Wu, H.C.; Yang, T.I. Electrospun Hydrophobic Polyaniline/Silk Fibroin Electrochromic Nanofibers with Low Electrical Resistance. *Polymers* **2020**, *12*, 2102. [CrossRef]

MDPI

St. Alban-Anlage 66

4052 Basel

Switzerland

Tel. +41 61 683 77 34

Fax +41 61 302 89 18

www.mdpi.com

Polymers Editorial Office

E-mail: polymers@mdpi.com

www.mdpi.com/journal/polymers

9 783036 545240